Agro-Advisory Services in Horticulture

NIPA® GENX ELECTRONIC RESOURCES & SOLUTIONS P. LTD.
New Delhi-110 034

About the Editors

Dr. Prabhat Kumar Shukla is presently working as Principal Scientist (Plant Pathology) at ICAR-CISH, Lucknow and ha also served as Assistant Professor at N.D. University of Agriculture and Technology in Kumarganj, Ayodhya. His specialization has been in disease survey, host parasite relationship, physiology of parasitism, factors influencing disease severity and nematode-fungus interaction and development of strategies for the management of important diseases of food grains, pulses, vegetables, medicinal and aromatic plants and fruit crops caused by plant parasitic nematodes and fungi. He has also worked on improvement in mushroom cultivation technologies and expansion of its cultivation through spawn supply and extension. He has 89 research papers, 23 book chapters, 2 manuals, 2 bulletins, 42 popular articles, and 18 extension folders to his credit. He has also received several awards, including the Young Scientist Award, the IPS Pathology Ratan Award, and the Best Scientist Award. He has also served as an examiner for several universities, as well as a reviewer/editor for Springer, APS, SIS, and other journals.

Dr. Tarun Adak presently working as Senior Scientist (ICAR-ARS) at ICAR-Central Institute of Subtropical Horticulture, Lucknow, Uttar Pradesh. He is a resource person for natural resource conservation and management in Horticultural crops. He is expert in agroadvisory services for issuance of real time weather based forecasting, soil and water management and mobile based advisory services in subtropical fruits. He is working on National Priorities. Worked sincerely and contributed significantly in NICRA project; working systematically and scientifically in the field of Science and management. Expert in providing key solutions in National Problems; disseminated technologies to farmers/growers/stakeholders across different states of India. Sincerely contributing to Skill India and Human Resource Development programmes of Govt. of India and devoted for Nation Building. He is Life Members/actively associated with number of National Societies/ Associations. He is authored and co-authored in research papers/articles/ bulletins in a number of peers reviewed National and International Journals. He is recipient of number of National Awards.

Dr. Gundappa distinguished entomologist and Senior Scientist at ICAR-NBAIR, Bengaluru. With over one and half decade of experience in agricultural research, his expertise spans around integrated pest management, insect biodiversity, and sustainable pest management practices. The main area of his research has been integrated pest management, development of decision support systems for IPM, Fruit Entomology, and climate change. Dr Gundappa's significant research contributions are the development of a forecasting model for major pests of mango and guava, validation of IPM technology for mango shoot gall psylla, guava bark eating caterpillar and mango thrips, development of species distribution models for quarantine pests of mango, and assessment of climate change on the insect pest distribution. He has also contributed to the validation of various technologies through the All India Coordinated Research Project on Fruits as well as AICRP on Biological Control of Crop Pests. He has developed more than 10 technologies and published extensively in peer-reviewed national and international journals. A recipient of multiple national and international accolades, he is committed to promoting environmentally friendly agricultural solutions for a resilient future.

Dr. Naresh Babu, Principal Scientist at ICAR-Central Institute for Subtropical Horticulture, Lucknow is a well known horticultural scientist who has served for over 25 years. He has obtained Master's and Ph.D. degree from Chandra Shekhar Azad University of Agriculture & Technology, Kanpur in 1990 and 1995. He has immensely contributed in hill and coastal horticulture for tribal people, nutrition, drip irrigation, canopy management, high density planting, organic farming, horticultural models, protected cultivation of vegetables and gender issues. He has developed orchards of various varieties of mango, guava, coconut, cashew nut and underutilized fruits. Presently he is working as resource person for crop diversification and dissemination of horticultural technologies covering farmers of different socio-economic condition. He has completed number of in-house; inter institutional and externally funded projects. He has good number of publications to his credits and life member of various scientific professional societies; organized several capacities building programmes for farming community and officials.

Agro-Advisory Services in Horticulture
Strategies for Climate-Resilient and Profitable Production

P K Shukla
Principal Scientist (Plant Pathology)
ICAR-Central Institute of Subtropical Horticulture
Lucknow, Uttar Pradesh - 226101, India

Tarun Adak
Senior Scientist
ICAR-Central Institute of Subtropical Horticulture
Lucknow, Uttar Pradesh - 226101, India

Gundappa
Senior Scientist (Entomology)
ICAR- National Bureau of Agricultural Insect Resources (NBAIR)
Bellary Road, Bengaluru – 560024, India

Naresh Babu
Principal Scientist (Horticulture)
ICAR-Central Institute of Subtropical Horticulture
Lucknow, Uttar Pradesh - 226101, India

NIPA® GENX ELECTRONIC RESOURCES & SOLUTIONS P. LTD.
New Delhi-110 034

nipa®

NIPA® GENX ELECTRONIC RESOURCES & SOLUTIONS P. LTD.

101,103, Vikas Surya Plaza, CU Block
L.S.C. Market, Pitam Pura, New Delhi-110 034
Ph : +91-11-43860225, Mob.: +91 9717133558, 9540816132
E-mail: newindiapublishingagency@gmail.com
Website: www.nipaersources.com

Print ISBN: 978-93-58872-98-9

ebook ISBN: 978-93-58879-48-3

Composed and Designed by NIPA®.

बिरसा कृषि विश्वविद्यालय

BIRSA AGRICULTURAL UNIVERSITY

राँची, झारखण्ड, भारत

Ranchi, Jharkhand, India

डॉ. सुनील चन्द्र दुबे
कुलपति
Dr. S.C. Dubey
Vice-Chancellor

Ref. No. : ....
Date: ....

Foreword

I am happy to know that the scientists of ICAR-CISH, Lucknow have compiled a book entitled Agro-advisory Services in Horticultural Crops. Best to my knowledge there are hardly any book is available on the subject. The content of the book indicates that all the aspects of the horticultural crops, including soil factors, water management, improved cultivars, canopy management, crop regulation practices and management of insect pests and diseases have been covered with reference to the climate change and day-to-day weather. The book is also providing information for advisories to the farmers on various aspects related to horticultural crops. Thus, it will be highly useful to the agencies issuing agro-advisory of horticultural crops on weekly basis. I hope that the book will be highly useful for the students, Academicians, Scientists, Stakeholders, Policy planners as well as farming community. I congratulate Dr. P.K. Shukla and his team for a very information compilation.

18/3/25

S.C Dubey

भा. कृ. अनु. प. - राष्ट्रीय कृषि कीट संसाधन ब्यूरो
(कृषि अनुसंधान एवं शिक्षा विभाग, कृषि एवं किसान कल्याण मंत्रालय, भारत सरकार)
डाक पेटिका संख्या 2491, एच. ए. फार्म पोस्ट, बेल्लारी रोड, हेब्बाल, बेंगलूरु – 560024,भारत

ICAR - NATIONAL BUREAU OF AGRICULTURAL INSECT RESOURCES
(Department of Agricultural Research and Education, Ministry of Agriculture & Farmers Welfare, Govt of India)
P. B. No. 2491, H. A. Farm Post, Bellary Road, Hebbal, Bengaluru - 560 024, INDIA
ISO 9001:2015 Certified Organisation

F. No. 12-4/ E024-25
Date: 20-03-2025

Dr. S.N. Sushil
Director

Foreword

Crop production is increasingly challenged by unpredictable climatic conditions and weather fluctuations. To support farmers in navigating these uncertainties, government agencies have been offering agro-advisory services. However, there has been a longstanding need to develop technological strategies that can mitigate the adverse effects of climate change on horticultural crops.

1 am pleased to note that the scientists at ICAR-Central Institute for Subtropical Horticulture, Lucknow, have taken a significant step in this direction by compiling a book based on past agro-advisory services offered to farmers. To my knowledge, no similar book exists that specifically addresses the unique agro-climatic conditions of Uttar Pradesh.

This book comprehensively covers all key aspects of horticultural crop management, including soil conditions, water management, improved cultivars, canopy management, crop regulation practices, and pest and disease management, all within the context of climate change and evolving weather patterns. Its content will be particularly valuable for agencies issuing weekly agro-advisories, as it provides a scientifically grounded foundation for such recommendations.

I am confident that this book will serve as an essential resource for students, academicians, scientists, stakeholders, policymakers, and the farming community. It is a timely and much- needed contribution to horticultural research and extension services. I sincerely hope it will gain widespread recognition and appreciation

Place: Bengaluru
Date: 20.03.2025

S.N. Sushil

Sardar Patel Outstanding ICAR Best Institute Award Winner
Website: htt://www.nbair.res.in E-mail: director.nbair@icar.gov.in
Phone:+91 (080) 23414220, 23417930

Preface

Food security is a serious concern and obviously national priority for any country. For a country like India, population trajectory is leading to most populous country by 2030 surpassing China. With this huge population to feed, role of scientific community to sustain food production and supply within the domestic market is a serious challenge. The target of United Nations' 2030 agenda to eradicate poverty and also to eradication of malnutrition, we shall achieve the target amicably. The role of scientific and farming community; policy planners and other stakeholders to support the on-going food security is praiseworthy. Climatic change has significant and tremendous affect on the food production system; be the annual crops or perennial crops all were affected by changing climatic scenarios across agro-climatic zones. In this direction, in order to achieve Sustainable Developmental Goals, role of real time based agroadvisory services plays significant role to sustain productivity via precise conservation and management of natural resources. Sincere efforts were put in sensitizing growers on timely adoption of management strategies to coup up the weather extremisms. Small and marginal farmers along with progressive growers were issued agro-advisory to follow and adopt for the betterment of their orchards/fields. Issuance of time bound nutrient management strategies, micronutrient application for enriching nutrient richness in fruits, weather based winter frost protection protocols, critical phenology based water application and insect and disease control measures needs to be strictly adopted at ground for ensuring orchards productivity vis-à-vis profitability. Mobile based advisory or decision support or expert based issuance or ICT based information diffusions are crucial in recent days to avoid weather vagaries and thereby fruit production management even in harsh soil and climatic condition. Horticulture sector plays major role in supporting the revenue generation either by domestic trade or foreign export. This sector has huge potential to support self sufficiency in a number of fruit crops grown and also scope to enhance the production by optimized use of natural resources. Environmental friendly crop protection measures are surely an answer to the safe production vs. farmers' earning. Thus, an attempt was made to compile recent information based on different aspects of farm advisories in horticultural crops widely cultivated across agro-ecosystems. The book contains various chapters contributed by the eminent scientists and specialists who have sincerely worked on the subject or devoted to horticulture based production systems. Hence, the latest publication has been brought out to sensitize and train human resources involved in extension activities, students, academicians, researchers and policy planners for efficient dissemination of technologies to growers for improving horticultural sustainability vis-à-vis national economy.

Editors

Contents

1

Importance of Agro-Advisory Services in Horticulture

T. Damodaran and P.K. Shukla

ICAR-Central Institute for Subtropical Horticulture, Rehmankhera P.O. Kakori, Lucknow - 226 101, Uttar Pradesh

Introduction

Crop production is weather dependent and almost every crop has been affected by the changing climate. The favourable weather may result bumper production sometimes or even complete destruction of crop may take place under severe weather aberrations. The number of events of sudden aberrations is increasing in a cropping period with the gradual climate change in the recent past. It has been recorded too severe in case of perennial crops. Ready to harvest crops have several times been completely destroyed due to thunderstorm or cyclone in coastal regions or even in main land. Thus, Indian agriculture is being impacted by the climatic changes, and its affect is very much visible now (Aggarwal, 2008). Horticulture sector is not escaped from its' ill affect and new challenges are coming for production (Singh *et al.*, 2021). The role of adaptation strategy has now been felt crucial (Babu *et al.*, 2021; Banerjee *et al.*, 2022) and importance of agro advisory not only protects the crops from such types of risks but also provide guidance to farmers about the right time for activities in crop husbandry (Adak *et al.*, 2018). Issuance for improving water productivity in water limited areas (Adak and Babu, 2024), micronutrient management for nutrient richness and enhanced yield (Adak *et al.*, 2023; Shukla *et al.*, 2020), rejuvenating unproductive orchards (Shukla *et al.*, 2021; Mishra *et al.*, 2007), disease (Misra *et al.*, 2013; Shukla *et al.*, 2015, 2019, 2020, 2021a,b; 2022; Kumari *et al.*, 2021) and insect pest (Gundappa *et al.*, 2016; 2018) management are becoming very crucial and essential for the benefit of growers. Moreover, for livelihood sustenance of small households of Uttar Pradesh, adoption of successful IFS model is essential which can be adopted at ground level (Meena *et al.*, 2022) and women's may get benefitted through integrated farming systems (Verma *et al.*, 2020). Thus, advisory in the form of benefiting people are highly welcomed at all levels of society.

Cursory Look at the Historical Events

Marginal and small farmers, as per the 2015-16 agriculture censuses, account for 86% of total farmers in India and are the most vulnerable to extreme weather events. Experts advocate protective measures for these vulnerable farmers. In 2021, Cyclone Tauktae and Cyclone Yaas wreaked havoc in several states in the first few months, especially in Odisha, West Bengal and Karnataka where lives and livelihoods were affected beyond redemption. In July, floods in Maharashtra damaged standing crops. This was followed by a 24 per cent nationwide rain deficit in August and 35 per cent excess rain in September. In October, heavy rains destroyed harvest-ready crops in many districts of Kerala. The devastation continued even towards the end of the year. In November, unprecedented rainfall caused huge loss of life and property in south Indian states of Andhra Pradesh, Tamil Nadu, Kerala, and Karnataka. Overall, India lost 5.04 million hectares (mha) crop area to cyclonic storm, flash floods, floods, landslides and cloudbursts till November 25, 2021, Narendra Singh Tomar, Union minister of agriculture and farmers' welfare, told the LokSabha November 30. Extreme weather events throughout the year have become the new normal. The decade spanning 2010-2019 was the most turbulent for disasters, Food and Agriculture Organization's (FAO) assessment showed. In India, around 36 m ha agricultural area was affected due to hydro-meteorological calamities, including heavy rain and floods since 2016 – 6.65 m ha in 2016, 5.08 m ha in 2017, 1.70 m ha in 2018, 11.42 m ha in 2019, 6.65 m ha in 2020 and 5.04 m ha in 2021. The above data has been aggregated from government replies in the parliament on crop loss and damage in recent years. This has led to repeated losses for farmers, especially small and marginal ones who comprise over 85 per cent of the total number of farmers in the country. They are being increasingly put to test, as climate change disturbs everything from sowing operations to harvest. There is no available national data on the monetary loss owing to damage of crops in 2021 but a central government assessment of the value of damaged crops for the years 2016 and 2017 can help throw some light.

In 2016, damage to crops on 6.65 mha area was estimated to be to the tune of Rs 4,052.72 crore, the Union government said in the LokSabha in 2020. If this is applied to the 36 m ha of crop area India has lost since 2016, the figure translates to a whopping Rs 29,939 crore loss for the farmers. However, this would only be an indicative figure as the 2017 evaluation of the crop damage suggests. In 2017, while the area of crop damage was lower compared to

2016 — 5.08 m ha the value of such damage was more than double – Rs 8,761. 39 crore. So, a lot also depends on the type of crops and the sowing season. For example, farmers suffer more losses if the crop loss happens during the Kharif season as most earn during that time. On the other hand, the state of crop insurance or compensation mechanisms is not desirable or efficient. Over half of all shocks to crop production are the result of extreme weather events, reinforcing concern about the vulnerability of arable systems to climatic and meteorological volatility, said FAO in its recent State of Food and Agriculture, 2021.

Reality of Price Rise and Role of Scientific Approaches for Sustained Production

It has been observed for decades that climate change seriously alters food prices. The average food prices (after adjusting for inflation) in the 11 months of 2021 were the highest in 46 years. The global food price rise was driven predominantly by wheat, which reported an increase due to drought and high temperature in major producing countries, including the United States and Canada. In 2021, as various trade reports showed, spring wheat production declined by 40 per cent in US due to drought and heat. Russia, the world's largest exporter of wheat, is estimated to harvest less due to unfavourable weather conditions this season. It has imposed a tax on wheat export to ensure ample stock for domestic consumption. Even in India, tomato prices hit the roof towards November end, in the wake of devastating rains in southern India. Among other reasons for such price hikes, erratic weather has become one of the prime causes in recent times. The need of the hour is focusing on developing and strengthening resilient capacities of especially smallholder farmers. The impact of weather on phenology of guava in Uttar Pradesh was assessed (Mishra *et al.*, 2014) and the usefulness of agromet advisory services in food production systems was thought off and suggested by Ramachandrappa *et al.* (2018). Even, Kumar *et al.* (2022) advocated for developing agromet advisory bulletins for the benefit of large scale food producers. Since, rainfall is a key issue in food production, its characteristics under changing climate should also be assessed to have knowledge of risk of crop damage (Singh and Rath, 2024). Drip fertigation in this direction played pivotal role to save water and also produce fruit (Thanki *et al.*, 2022). Scientific approaches for producing quality fruits through nutrient management is essential (Adak *et al.*, 2019) and adoption of organic farming is the realization now a days (Shukla and Adak, 2024). Quantifying the soil health indicators through scientific approaches was thus realized as mandatory protocol (Adak and Pandey, 2020) and suggested for management practices to be followed at ground level for ensuring sustainability (Adak *et al.,* 2022). Even, intercropping of turmeric in the interspaces of mango orchards benefitted growers with greater revenue generation (Shukla and Adak, 2020) and Pepino cultivation in subtropical climates may also enhanced growers income with less investment (Kumar *et al.,* 2017b). Incorporation of green manuring and nutrient management ensured sustainability in Dashehari mango (Kumar *et al.,* 2017a). Management of insect pests (Gundappa *et al.,* 2015, 2016; 2018) and diseases (Misra *et al.,* 2013; Shukla *et al.,* 2015, 2019, 2020, 2021a,b; 2022; Kumari *et al.*, 2021) of fruit crops under subtropical conditions are economically important for saving the fruits and earning. Orchard management through scientific technologies is thus very much important and advisory services to growers are helpful for profits and quality produces realization (Kumar *et al.,* 2016; Babu *et al.,* 2022). The thrust of fruit security through resource conservation and management was thus should be played seriously and meticulously (Adak and Mishra, 2024).

Realization of Agro-meteorological Services Benefiting Farmers

The Ministry of Earth Sciences (MoES) in collaboration with the Indian Meteorological Department (IMD) has developed a good network of agro-meteorological advisory services to support Indian farmers. The crop production is influenced by large number of components, including soil, seed, planting material, horticultural operations, nutrient management, crop production and protection techniques, weather conditions, etc. Since, subtropical fruit crops are mostly perennial, their development, flowering and fruiting are governed by the seasons. Here weather stands as the foremost determinant in their growth and productivity. AAS contribute to collect and organize climate/weather data, soil and crop information and to integrate them with weather forecast to assist farmers in taking management decisions. These services encompass a wide range of offerings, including weather forecasts (such as Nowcast, medium-range, and long-range weather forecasts), weather-based crop advisories, fundamental cultural practices for crop production, short message services, and the dissemination of Agrometeorological Advisory Bulletins (AABs) (SOP, IMD, 2020).

The emerging capacity to provide timely, skillful weather forecasts offers the potential to reduce vulnerability to vagaries of weather (Hansen, 2002). Agriculturally relevant forecast is not only useful for efficient management of farm inputs but also leads to precise impact assessment (Gadgil, 1989 and Anon., 2002). A reliable forecast on weather has been basis for the agro advisories in relation to the crop stages. This work has been done as part of the NICRA project operated in different agro-ecosystems in India with prime objective to minimize the risks in crop production through implementation of real time contingency practices in relation to prevailing climatic situations. The study showed the extent of benefits by adopting ago advisory services for aberrant weather condition based on weather forecast. The project "National Innovations is Climate Resilient Agriculture (NICRA)" funded by ICAR and coordinated by Central Research Institute for Dryland Agriculture (CRIDA) AICRPDA, UAS, Bengaluru has been working on real time contingency plans at Chikkamaranahalli cluster (Chikkamaranahalli, Chikkamaranahalli colony, Chikkaputtayyanapalya, Mudalapalya and Hosapalya), Nelamangalataluk, Bengaluru Rural district since 2011. The socio economic study indicated that the majority of the farmers in the study village were marginal and small (144 and 48, respectively) while seven farmers had medium land holdings and only two farmers had large holdings. The total area in the cluster consisted of 579 ha out of which the total cultivated area was 409.2 ha (irrigated 18.0 ha and rainfed 367.4 ha, followed by garden land and uncultivable land is 23.8 ha and 169.8 ha, respectively). The study was undertaken on adaptation of weather based agromet advisories and economic impact of agromet advisory services and the same were compared to with farmers who have not followed. The economic impact studies indicated that there was considerable benefit to farmers who adopted the agromet advisories. The per cent gain in income due to adoption of suggested contingency cropping systems ranged from 22 to 397% over traditional cropping systems. The yield increase in cropping systems viz., finger millet + pigeonpea, groundnut + pigeonpea, pigeonpea + cowpea and pigeonpea + field bean was to the tune of 1040, 741, 481 and 594 kg/ha, respectively. Intercropping is a preparedness type of contingency for aberrant weather condition as the farmer can realize income from one crop if other fails. Hence, intercropping is additionally considered besides other real time contingency plan. Transplanting of finger millet registered higher yield as contingency measure for early season drought. Rajegowda *et al.* (2008) reported that, in the Eastern dry zone of Karnataka the farmers who have adopted the agromet advisories have realized an average economic benefit of 31.4, 24.7, 16.2 and 20.6% in finger millet, pigeonpea, field bean and tomato, respectively. Hence, it can be seen that the agro advisory information communicated to farmers in NICRA villages benefited the farmers. Majority of farmers who followed the weather related agromet advisories expressed that they were able to reduce the crop losses. Profit to farmers due to adoption of agromet advisory services was enhanced due to adoption of weather based crop management practices which include timely land preparation and sowing, adoption of recommended seed rate and suitable varieties, timely weeding, opening of *in-situ* moisture conservation furrow between paired rows of pigeonpea to conserve rainwater and maintain the soil moisture, pesticide application, irrigation and harvesting which helped in realizing higher yield of finger millet and groundnut cropping system (Vashisth *et al.,* 2013). Wide array of finger millet varieties with different duration and season has been developed and released for cultivation. But farmers are growing long/medium/short duration varieties irrespective of the sowing time. Long duration (120-125 days) variety MR-1 for July sowing and medium duration (110-115 days) variety GPU-28 for August 1st fortnight sowing was recommended as real time agro advisory. Due to the regular onset of monsoon during 2011, 2013 and 2014, long duration variety MR-1 sown during the month of July recorded higher grain yield, net returns and B:C ratio compared to medium duration variety. Delayed onset of monsoon during 2012, the agro advisory messages suggested the farmers to adopt medium duration variety (GPU-28) for late sowing in the month of August and repeated interculture operations during dryspells, besides balanced nutrient application which mitigated the drought during crop growth season. The AAS farmers who adopted advisories and sown GPU-28 have realized higher grain yield, net returns and B: C ratio (1720 kg/ha, Rs. 18096/ha and 2.31, respectively), while the non AAS farmers adopted long duration variety (MR-1) even under the late onset of monsoon and incurred huge losses. Ramachandrappa *et al.* (2010) also reported the similar results. During 2015, after a normal onset of monsoon, there was dry spell during panicle initiation stage for long duration variety (MR-1). However, rainfall (197.6 mm) during November month facilitated in increased yield of medium duration variety GPU-28.

Frequency and coverage of weather forecast used by the farmers was revealed from the study that most of the farmers gave more importance to forecast of rainfall over temperature, wind speed and cloud cover. Wind speed and cloud cover play an important role during infestation of pest and disease. About 80% farmers were keen to have bi-weekly rainfall forecast of rainfall to carryout various agriculture operations. Forecast on wind and cloud cover by 62 and 60% of farmers, respectively at bi weekly intervals. About 54% of the farmers were not interested to obtain the forecast of temperature. Majority of the farmers opined that the weather bulletin was received by them on time. A large number of farmers were happy with the overall usefulness of the AAS bulletins issued. Similar results of timely availability of clear information through agroadvisory services were reported by Ram Singh *et al.* (2015). The qualities of forecast on agro advisory services were studied in the domain villages and 46% of the farmers opined that the forecast on weather related parameters and harvest and post harvest measures were good while forecast on field operations as very good. Farmers appreciated weather based agro advisory service and utilized the advice in scheduling of irrigation, application of chemical fertilizers and in deciding best time for taking control measures (spraying) for pest and diseases. Survey conducted at mid hill-region of Himachal Pradesh indicated that 38% of farmers rated agromet advisories as excellent and good by 29% of farmers (Rana *et al.*, 2005).

Conclusion

The observations on the application of agromet advisory, based on current and forecasted weather has been found highly useful in protection of crops and in enhancing the production and income. Farmers received weather forecast based agro-advisories, including selection of suitable crops and varieties, and farm activity advices on real time basis have been benefited. The increased yield level and reduced cost of cultivation led to increased profit.

References

Adak, T., Kumar K. and Singh V.K. 2019. Assessing micronutrient management and fertilizer doses on soil and foliar properties and yield in Dashehari mango grown orchard soils of subtropical region. Tropical Plant Research, 6(3): 417-423.

Adak, T., Babu, N. and Singh V. K. 2022. Mango Biodiversity Conservation: Management Needs for Better Sustainability. In: Soil and Water Conservation Bulletin (Mandal, D., Kaushal, R., Kumar, G., Singh, R. J. and Roy, T. (eds.), Indian Association of Soil and Water Conservationists, Dehradun, Uttarakhand, No. 7: 29-36.

Adak, T., Kumar K., Ganeshamurthy A. N. and Pandey G. 2023. Yield sustainability through micronutrient management in guava. Journal of Horticultural Sciences, 18(2): 345-350.

Adak, T., Shukla, P. K., Singh, A. K., Gundappa and Chandra, S. 2018. A new approach to reach farmers through agroadvisory services for better orchard management. Kahaar, 5(3): 31-36.

Adak, T. and Babu, N. 2024. Assessing Evapotranspiration and Water Productivity in mango cv. Amrapali under Subtropical Climate. Journal of Agricultural Physics, 24 (1): 81-94.

Adak, T. and Mishra, D. 2024. Scientific efforts for fruit security: Yield gap analysis, challenges and future perspectives. Current Advances in Agricultural Sciences, 16(2): 43-54.

Adak, T. and Pandey, G. 2020. Soil enzymatic activity as a soil health indicator for mango orchards under subtropical region. Journal of Soil and Water Conservation 19(4): 364-369.

Aggarwal, P. K. 2008. Global climate change and Indian agriculture: impacts, adaptation and mitigation. Indian Journal of Agricultural Sciences, 78 (10): 911-919.

Anonymous. 2002. Background paper for 11th Annual Review Meeting of National Centre for Medium Range Weather Forecasting (NCMRWF) New Delhi. In: Status of economic impact. Pp. 1-3.

Babu, N., Adak, T., Arvind, K. and Singh, V. K. 2021. Scientific analysis of drought assessment and its management strategies for food security. Current Advances in Agricultural Sciences (An International Journal), 13(1): 1-6. DOI: 10. 5958/2394-4471. 2021. 00001. 0.

Babu, N., Adak, T., Chandra, S. and Kumar, A. 2020. Crop Diversification in Fruit Belts - Sole Way to Enhance Farmers Income. In: Soil and Water Conservation Bulletin 5, Dogra, P., Sharma, N. K., Mandal, D., Kumar, G., Rawat, I., Bishnoi, R., Kar, S. K. and Kumar, S. (Eds), Indian Association of Soil and Water Conservationists, Dehradun, Uttarakhand, 2020, pp 89-94.

Babu, N., Adak, T. and Kumar, A. 2022. Horticulture intervention for improving productivity and profitability of small and marginal farmers. Journal of Natural Resource Conservation and Management, 3(1): 49-57.

Banerjee, S., Sarmah, K., Mukherjee,. A, Sattar, A. and Bandopadhyay, P. 2022. Effect of projected climate scenarios on the yields of potato crop and agronomic adaptation options as evaluated by crop growth model. Mausam, 73 (1): 71-78.

Gadgil, S. 1989. Monsoon variability and its relationship with agricultural strategies. Paper presented at International symposium on climate change variability and food security in developing countries. Feb 5-7, 1987, New Delhi, India, pp. 249-267.

Gundappa, Adak, T. and Rajkumar, B. 2015. Variability of thermal indices and peak mango hopper incidence under subtropical conditions. Journal of Agricultural Physics, 15(2):150-156.

Gundappa, Adak, T. and Shukla, P. K. 2016. Appraisal of thrips population dynamics in mango using weather based indices. Vegetos, 29(3):138-145.

Gundappa, Adak, T. and Shukla, P. K. 2018. Population dynamics of mango mealybug (Drosicha mangiferae) (Margorididae: Hemiptera) and its relation with weather parameters in subtropical climatic conditions. Indian Journal of Agricultural Sciences, 88 (6): 865-870.

Hansen, J. W. 2002. Realizing the potential benefits of climate prediction to agriculture: Issues, approaches and challenges. Agricultural Systems, 74: 329-330.

Kumar, A, Adak T. and Rajan S. 2017b. Pepino (Solanum muricatum Ait.): A potential future crop for subtropics. Tropical Plant Research 4(3): 514-517.

Kumar, A., Pandey, A., Singh S. K., Prasad, L. and Ghosh, K. 2022. Tools for preparation of agromet advisory bulletins. AMFU Roorkee, WRD&M Department, IIT Roorkee, pp 1.

Kumar, K., Adak T. and Singh V.K., 2017a. Green manuring and nutrient management impacting soil properties and sustainability of mango orchard. Journal of Soil and Water Conservation, 16(1): 72-78.

Kumar, K., Adak, T., Lal, B., Shukla, P. K., Balaji Rajkumar, Chandra, S. and, Singh, V. K. 2016. Technological intervention in sustaining mango productivity in a resource poor farmers' field of Uttar Pradesh. GERF Bulletin of Biosciences, 7(2):6-11.

Kumari Nidhi, Sneha Chaudhary, Shiwali Dhiman, P. K. Shukla and Gundappa (2021). Biological management of horticultural crop diseases using phyllosphere-, rhizosphere-, and endosphere-derived biocontrol agents. In: Innovative Approaches in Diagnosis and Management of Crop Diseases, Vol. 3 Nanomolecules and biocontrol agents, Eds. R. K. Singh and Gopala, AAP Inc, Canada, pp. 179-200.

Meena, L. R., Malik, S., Mishra, D., Nath, A., Prusty, A. K., Chandra Bhanu, Kumar, S., Meena, A. L., Subash, N., Meena, S. R. and Singh, A. 2022. Development of a successful IFS model for livelihood sustenance of small households of Uttar Pradesh. The Indian Journal of Agricultural Sciences, 92(10): 1175-1180.

Mishra, D, Pandey D, Mishra R and Pathak R K, 2007. Performance of improved aonla cultivars during top-working of senile trees. Indian Journal of Horticulture, 64(4): 396-398.

Mishra, D, Shukla S K, Ravishankar H and Adak, T. 2014. Impact of weather on phenology of guava in Uttar Pradesh: A cursory analysis. Current Advances in Agricultural Sciences, 6(1): 74-75.

Misra, A. K., Shukla, P. K. and Pandey, B. K. (2013). Diseases of mango. In: Diseases of fruit crops (A. K. Misra, P. Chowdappa, Pratibha Sharma and R. K. Khetrapal Eds.), IPS, New Delhi Publication, pp. 277-335.

Rajegowda, M. B., Janardhanagowda N. A., Jagadeesha N and Ravindrababu B. T. 2008. Influence of agromet advisory services on economic impact of crops. Journal of Agrometeorology, 10: 215-218.

Ramachandrappa, B. K., Balakrishna Reddy P. C, Mariraju H. and Ashok Kumar H. P. 2010. Impact of operational research project for dryland agriculture. Indian Journal of Dryland Agricultural Research and Development, 25 (1): 92-98.

Ramachandrappa, B. K., Thimmegowda, M. N., Krishnamurthy, R., Srikanth Babu, P. N., Savitha, M. S., Ch. Srinivasarao, Gopinath, K. A. and G. Ravindra Chary. 2018. Usefulness and impact of agromet advisory services in eastern dry zone of Karnataka. Indian J. Dryland Agric. Res. & Dev. 33(1): 32-36.

Ram Singh, Willey Syiem, Feroze, S. M., Devarani L, Lala Ray I. P., Singh A. K., Singh N. J. and Anurag T. S. 2015. Impact assessment of mobile based agro-advisory: A Case study of tribal farmers of Ri-Bhoi district of Meghalaya. Agricultural Economics Research Review, 28: 183-187.

Rana, R. S., Prasad R and Kumar S. 2005. Reliability of medium range weather forecast in mid hill region of Himachal Pradesh. Journal of Agrometeorology, 7 (2): 297-303.

Shukla, P. K., Adak, T., Misra, A. K. and Singh, A. 2016b. Appraisal of shoulder browning disease of mango (mangifera indica L.) caused by Fungus complex in subtropical region of India. J. Mycol. Pl. Pathol. 46(1): 38-46.

Shukla, P. K., Adak, T. and Gundappa. 2016a. Seasonal dynamics of powdery mildew of mango and its management under subtropics. GERF Bull Biosc, 7(1):21-25.

Shukla, P. K., Adak, T. and Gundappa. 2017. Anthracnose disease dynamics of mango orchards in relation to humid thermal ratio under subtropical climatic condition. Journal of Agrometeorology 19(1): 56-61.

Shukla, P. K., Fatima, T., Gundappa and Nidhi Kumari (2020). Non-chemical Management of Postharvest Diseases of Mango. In: Bio-management of Postharvest Diseases and Mycotoxigenic Fungi Eds. Neeta Sharma and Avantina S. Bhandari, CRC Press, Taylor & Francis Group, LLC. ISBN: [978-1-4987-9731-3] (hbk) ISBN: [978-1-003-08922-3] (ebk) pp. 89-102.

Shukla, P. K., Fatima, T. and Misra, A. K. (2019). Recent trends in management of major diseases of mango. In: Recent Trends in Plant Disease Management, eds. Sheela, Chourasia, H. K., Sahu, K. K., Jha, R. N. and Prasad, G., Today & Tomorrow's Printers and Publishers, New Delhi, ISBN 10: 81-7019-644-5/ ISBN 9788170196440, pp. 35-68.

Shukla, P. K., Nidhi Kumari and Tahseen Fatima (2022). Major Diseases of Guava, Bael and Aonla. In: Diseases of fruits and plantation crops and their sustainable management. Eds. M. R. Khan and Ziaul Haque. Nova Science Publ., Inc., New York, pp. 307-328.

Shukla, S. K., Adak, T. and Singh, V. K. 2020. Evaluation of boron nutrition in enhancing productivity and quality of mango (Mangifera indica L.) cultivar Mallika under subtropical climatic conditions. Current Advances in Agricultural Sciences, 12(1): 33-36.

Shukla, S. K. and Adak, T. 2020. Intercropping of turmeric for revenue generation from interspaces in mango orchards. Kahaar 7(1-2): 41-43.

Shukla, S. K. and Adak, T. 2024. Assessment of soil fertility vis-à-vis yield in mango cvs Dashehari and Lucknow Safeda under organic farming. Journal of Eco-friendly Agriculture, 19(2): 247-252.

Shukla, SK, Mishra D Pandey G and Rajan S. 2021. Rejuvenating old and unproductive mango orchards for enhanced farmers' income. Indian Horticulture, 66(4): 32-34.

Shukla P. K. (2021). Advances in the effective management of major diseases of mango and guava. In: Technology Strides in Plant Health Management, Eds. N. K. Bharat and H. R. Gautam, Neoti Book Agency Pvt. Ltd., New Delhi, pp. 336-359. ISBN 9788194566403.

Shukla P. K., Fatima, T., Gundappa and Nidhi Kumari (2021). Postharvest diseases of mango and their management. In: Postharvest handling and diseases of horticultural produces. Eds. Dinesh Singh, R. R. Sharma, V. Devappa and D. Kamil, CRC Press, New York, pp. 189-199.

Shukla P. K., Misra A. K. and Gundappa (2015). Mango decline and wilt may be a serious threat for future in India. Proceedings of the 36th Annual Conference and National Symposium on "Challenges and management approaches for crop diseases of national importance-Status and prospects" organized by ISMPP, Udaipur held at Madurai, India during 12-15 February 2015, pp. 318-325.

Singh, A. K., Sanakaran, M. and Murthy, B. N. S. 2021. Horticulture for food, nutritional and socio-economic security in India. Indian Horticulture, 9-10: 1-9.

Singh, M. and Rath, B. S. 2024. Rainfall characteristics under changing climate in Sundergarh district of Odisha, Mausam, 75 (4): 993-1008.

Standard Operating Procedure for Agromet Advisory Services (2020). India Meteorological Department Ministry of Earth Sciences, New Delhi. https://mausam. imd. gov. in/imd_latest/contents/pdf/gkms_sop. pdf.

Thanki, D. M., Patel, K. D. and Kumar, S. 2022. Influence of fertigation schedule on yield of dragon fruit (Hylocereus polyrhizus Britton & Rose). The Pharma Innovation Journal, 11(9): 518-521.

Vashisth, A., Singh R, Das D. K. and Baloda R. 2013. Weather based agromet advisories for enhancing the production and income of the farmers under changing climate scenario. International Journal of Agriculture and Food Science Technology, 9: 847-850.

Verma, N., Mishra, D., Kumar, S., Singh, S. P. and Panwar, A. S. 2020. Income generation for women farmers through integrated farming systems: A successful case from Mandora village. Indian Farming, 70(2): 27-29.

2

Recent Scientific Efforts in Agro-Advisory Services in Fruit Crops for the Benefits of Subtropical Fruit Growers

Tarun Adak, P.K. Shukla, Gundappa and Dushyant Mishra

ICAR- Central Institute for Subtropical Horticulture, Rehmankhera Lucknow-226101, Uttar Pradesh, India

Introduction

World population is increasing and currently stands at 8.2 Billion as on 2025. Food security of population of any country is becoming a challenging task and the entire population depends on the available national resources. The importance of scientific community in securing foods for everybody was felt necessary and it was expressed that there is a huge herculean task to ensure food for everybody. Farming community and fruit growers across the globe were engaged in serious efforts to produce fruits of good quality. The necessity to develop advanced technologies for nutritious foods was thus given national importance not only for food security but also of nutritional security. The sincere scientific efforts to develop recent innovative technologies to support fruit production even at harsh soils and environmental conditions are praiseworthy. Farmers, stakeholders, policy planners and technology developers have been devoted to achieve the target of Sustainable Developmental Goals. The target of United Nation's 2030 to eradicate poverty and eradication of malnutrition should be achieved sustainably. People across nations are working devotedly to achieve the target. However, climatic changes have been significantly affecting the food production to food security across nations. Decadal information suggested the adverse impact of climates on the fruit production. Recent changes in weather extremisms had also affect harshly on the food production to supply chain. Thus, the role of scientific community to coup up the environmental hazards and to sustain fruit production becomes a reality. In this context, agroadvisory services plays significant role to support farming community and fruit growers to adopt time bound managemental practices. The ultimate aim of these services is to improve the orchard soil and tree health condition, enhance factor productivity *vis-à-vis* profitability and precise use of available natural resources. Quantification of variability in environmental indicators, soil and tree interactions were felt necessary for formulating effective policy plan (Malhotra, 2017). Adak *et al.* (2022) advocated for foliage spraying of 1 per cent zinc sulphate on mango *cv.* Dashehari to enhance yield and nutrient richness in fruit pulp.

The aim of issuance of advisory services was also to attain higher sustainability through achieving greater sustainable yield index. In this context, growers were advocated to adopt micronutrients in their nutrient management practices effectively along with NPK and FYM application. Inclusion of green manuring and bio-fertilizers were also encouraged as all these inputs had positive impacts over the years and enhanced sustainable yield index in bearing mango orchards (Adak *et al.,* 2020a). To sustain soil biological health, systematic evaluation of biological parameters from farmers orchards were implemented and it was found that soil biological health was designated as poor, low and in some cases medium across mango orchards of subtropical region (Adak *et al*., 2018). Therefore, to increase the health status, ICAR-CISH developed Bio-enhancer was recommended to growers for immediate use. Sometimes severe drought like condition arises with extremely high temperatures and absence of rainfall; very high reference evapotranspiration made dried situation wherein precise management practices were advocated to growers for mitigating weather vulnerabilities (Babu *et al.,* 2021). BBCH scale based mealybug attack in mango as a function of weather indicators were scientifically analysis based on which farmers were issued effective agroadvisory to control mealybug infestation in mango orchards across regions (Gundappa *et al.,* 2018). Apart from crop protection measures, farmers were also encouraged to grow turmeric in interspaces of mango orchards to generate good revenue (Shukla and Adak, 2020) and to further improving income, strawberry cultivation even in subtropical climate was demonstrated and adopted at farmers field (Kumar *et al.,* 2016). As soil and tree health management are necessary in arid, semi-arid, rain fed, coastal, Himalayan, Hilly and other agro ecologies, integrated nutrient practices in aonla was advocated to growers for better yield, quality and soil health management (Singh *et al.,* 2015) and imbalances of micronutrients were removed following site-specific micronutrient practices in sapota (Rashmi *et al.,* 2020). In all other fruit crops, flowers efficient resource conservation and management practices were advocated to growers for the betterment of overall system productivity and sustainability.

Agro-advisory Services for Optimizing Natural Resources

Now-a-days, a serious challenge of scientific community ahead is due to shrinking natural resources. Therefore, optimization of resources, its conservation and management becomes very critical and crucial (Fig 1 to 2). The performance of orchards having similar fruit cultivar varies under different climates and different soils. Role of site specific nutrient management was thus given due importance. Micronutrient practices for both soil and tree foliage is essentially important to improve nutrient richness in fruit pulps. Weather based real time water application in critical reproductive stages are very much crucial and farmers were advocated to follow advanced protocols based on recent weather data. Day to day weather analysis and weekly based forecasting systems were thus very much beneficial for growers to avoid fruit damage and productivity losses. Low cost fruit production technologies were also diffused to growers for their adoption at field level and to get more profits. The technologies suitable for progressive farmers were adopted at ground level and positive implications of agroadvisory services were evolved. Benefits of farmers in terms of cost savings, saving of resources were ensured (Table 1). Small and marginal farmers were trained, sensitized and awareness created to enhance perception of positive role of orchard management. The importance of rejuvenation, incorporation of organic fertilizers, adoption of crop diversification, fertigation, low cost strawberry production, precise mango ad guava production, apple ber, apple and other fruits in non-traditional area was given die focus to enhance profitability per unit area (Fig 3 to 4). Therefore, positive response and significant outcomes of agroadvisory services adopted by fruit growers of subtropical and other agro-climatic zone was achieved.

Fig. 1: Time bound water application for ensuring guava fruit production

Fig. 2: Need for efficient micronutrient practices for enhancing Mango cv Dashehari and Mallika production

Table 1: Effective resource conservation and management advisory for growers benefit

Horticulture system	Technological advices	References
Guava	For improving yield and quality, use of 0.75 to 1.0 per cent zinc sulphate spraying at critical fruit setting to developmental stages are advocated	Adak *et al.* (2020c)
Mango	Application of NPK (1000g N, 500g P_2O_5: 500g K_2O/year/tree) along with 85 gram boron metasolate for higher yield and quality in Mallika mango	Shukla *et al.* (2020)
Dragon fruit	Use of 400 gram calcium nitrate per dragon fruit hill per year for enhanced yield and quality	Sahu *et al.* (2022)
Papaya	Water application at 40 per cent evaporation replenishment through shifting of one emitter at 15 days intervals for greater yield and water saving	Manjunath *et al.* (2017)

Adak *et al.* (2019) advocated to growers for adopting drip irrigation based system for irrigation water practices for not only saving of ground water but also to enhance yield and water use efficiency in subtropical fruits. For improving sustainable yield index in guava, Adak *et al.* (2020b) recommended grower should apply 120, 60, 50g NPK/tree/year of age along with foliar application of Zn, B, Mn and Cu or else 120, 60, 50g NPK/tree/year of age along with soil application of Zn, B, Mn and Cu. Incorporation of both organic [5kg FYM along with *Azotobacter* + PSM + *Trichoderma harzianum* + organic mulching (10 cm thick)] and inorganic (120, 60, 50g NPK/tree/year of age). In case of high density system in mango *cv.* Amrapali, Sharma *et al.* (2016) suggested to apply 415:130:360 NPK g $tree^{-1}$ along with 25 kg vermicompost, 2.5 kg oil cake, 100g *Azotobacter*, 100g VAM, 100g TV and 100g PSB for higher fruit set pea stage, fruit retention, number of fruit tree^{-1} and B:C ratio.

Fig. 3: Issuance for rejuvenation of unproductive orchards for future fruit production

In order to improve the total sugar content in pomegranate, potassium of 3 g/L (potassium metalosate) and Boron of 3 g/L should be sprayed in orchards (Tehranifar and Tabar, 2009). Dalal *et al.* (2017) suggested Kinnow mandarin growers that to increase yield in Kinnow mandarin, split doses of fertilizers were adopted. Nitrogen was applied three times, 40 per cent in February and 30 per cent each in April and July. Phosphorus was applied two times, 50 per cent each in February and April. Potash was applied three times, 40 per cent each in February and April and 20 per cent in September. The applied fertilizers had reduced pre-harvest fruit drop (21.16 per cent), increased number of fruits (10.19 per cent) and increase in fruit yield about 10.29 per cent in Kinnow. Growers were also advised to spray borax and zinc at 0.2 per cent concentration of each for higher fruit production in respect of yield (7.2 to 8.5 kg), bunch weight (233 to 289 g) and ten berry weight (29.1 to 29.2 g) but also improved the fruit quality in respect of higher TSS/acid ratio (35.4 to 38.7) in four years old Arka Neelamani grape (Pal and Ghosh, 2018). Moreover, farmers were also advised to spray 0.4 per cent calcium chloride and 0.2 per cent boric acid in increasing quality TSS (11.99°B), ascorbic acid (252.4 mg/100 g) and total sugars (7.51 per cent) in guava *cv.* Pant Prabhat (Kumar *et al.*, 2017). All these findings may be incorporated in agro-advisory to growers for yield and quality enhancement along with profitability point of view.

Fig. 4: Agro-advisory to growers for newly established tree protection and water application during peak summers

Efficient Agro-advisory Services for Rationalizing Crop Protection

The role of agroadvisory services to farmers for precise crop protection practices was havoc and significant at each stages of production cycles. The efficient ways of reduction in number of pesticides spray through decision making on the basis of advisory, adoption of good agricultural practices, time bound insect and diseases management are some of the positive outcomes. It was recorded that sincere scientific efforts have brought down the pesticide load by adoption of fruit fly trap and saving of around Rs. 10,000/ha cost of growers. Real time information on infestation of leaf Webber, mealy bug, thrips, hoppers, fruit borers, semi-looper, fruit fly are encouraged to growers with positive note. Growers' get benefited greatly due to adoption of weather based advisory services even at micro level. Crop performances have improved and better quality fruit production was ensured. The cost of cultivation of fruit was lowering down with adoption of alternative means of pest management. The issuance of advisory to reduce the sucking pest is praiseworthy. Training of the stakeholders, sensitization and trained man power was thus developing to ensure control of any pest attack. The problem of emergence of new insects due to environmental hazards was also felt and farmers were sensitized from time to time through mobile services, media reports in local languages. The approach for efficient crop protection was different for newly already bearing trees. It was much more relevant to put sincere effort to sustain fruit production via integrated approaches across agro-climatic zones.

The environmental vulnerabilities need to be quantified, predicted for developing future mitigating strategies. Effective implementation of these climate smart strategies would certainly benefit people across country. Rozante *et al.* (2020) scientifically developed frost prediction index and it would be beneficial to growers. Forecasting models were also effective and growers need to follow weather based control practices in leaf webber incidences in mango orchards (Gundappa *et al.*, 2016b) and also based on humid thermal ratio, mango thrips dynamics may be predicted. Control of thrips infestation is essential and mandatory to avoid mango fruit loss (Gundappa *et al.*, 2016c). Therefore scientific efforts were put in place to improve the orchard health status and to provide better remunerative to farmers' across socio-economic condition (Fig 5 to 6).

Fig. 5: Need for efficient guava tree health management and insect controls

Fig. 6: Recent farmers' interaction with scientists and kisan gosthi at 2021 to 2024 for crop protection

The powdery mildew and blossom blight diseases are endemic to epidemic in mango growing areas. There incidence and severity are weather dependent. There was a time when farmers were spraying fungicides at 7 days interval during whole flowering period; however, with the scientific analysis of disease triangle, the critical stage was identified and recommendation was made for 1 or maximum 2 spray for the management of these diseases (Shukla *et al.*, 2016a, b, c). Shoulder browning disease of mango fruits has also been a serious issue in Northern India; where fruits lying on trees during monsoon period becomes black and prone to severe post harvest rot. The issue was addressed and control measures were worked out; and extended to farmers with reference to weather predictions (Shukla *et al.*, 2016; 2018). Mango anthracnose is a serious disease of mango, infecting all tender parts of plants and causing fruit rot at ripening. It was also addressed in relation to weather and crop phenology; and management strategies were suggested (Shukla *et al.*, 2017). Mango wilt has also been gradually became a serious constraint in mango production. Technology was developed for its management after thorough studies, and included in the agro-advisory (Shukla *et al.*, 2018, 2019, 2020a,b). Root-knot nematode infestation in guava orchards and nurseries was surveyed, control measures were worked out and extended to farmers through advisory services (Khan *et al.*, 2019; Shukla *et al.*, 2023). Another severe constraint of wilt of banana was addressed and control measures were developed and extended to farmers (Damodaran *et al.*, 2019; Nidhi *et al.*, 2023). All those burning problems were picked from farmers orchards and in minimum possible period research was completed to help the farmers.

Future Scope for Advisory Services Across Nation

The perception of agroadvisory services to farmers was positive and growers easily accepted and adopted scientific advisories at their field/orchards. In near future, ICT based system, mobile apps, would be much more effective in diffusion of technologies. The importance lies to control sudden outbreak of any insects in orchards and their timely controls. Low cost technologies need to be evolved in controlling pest load and the menace is to be reduced by quality pesticide. Bio-formulations may act as potential role to contain pest menace. Orchardists should follow use of newly developed effective alternative means of pest and disease control to contain pest load. Farmers-scientist interactions for time to time should be followed in order to not only disseminate technologies but also enhance confidence among farming community. Thus, role of digital agroadvisory services should be ensured at all level to prosper farming community and thereby improving national economy.

References

Adak, T., Kumar, K., Pandey, G. and Singh, V. K. (2020c). Response of ZnSO4 through foliar nutrition on nutrient status, yield performance and fruit quality of guava cv. Shewta. Current Advances in Agricultural Sciences 12(2): 93-97.

Adak, T., Kumar, K. and Shukla, S. K. and Pandey, G. (2020b). Improving sustainable yield index in guava (Psidium guajava) through organic and inorganic inputs. Indian Journal of Agricultural Sciences 90 (7): 1267-1270.

Adak, T., Kumar K., Singh V. K. and Ganeshamurthy A. N. (2022). Response of Dashehari mango to different Zn levels on yield and pulp nutrient contents grown on sandy loam soils of Lucknow J. Hortl. Sci. 17(1): 103-109.

Adak, T., Pandey, G. and Kumar, K. (2020a). Improving Sustainable Yield Index in Mango through Integrated Nutrient Management. In: Soil and Water Conservation Bulletin 5, Dogra, P., Sharma, N. K., Mandal, D., Kumar, G., Rawat, I., Bishnoi, R., Kar, S. K. and Kumar, S. (eds), Indian Association of Soil and Water Conservationists, Dehradun, Uttarakhand, 2020, pp 40-47.

Adak, T., Pandey, G. and Rajan, S. (2019). Advanced knowledge on soil and water conservation measures in fruit orchards. Journal of Agricultural Physics 19(1): 35-45.

Adak, T., Shukla, P. K. and Gundappa. (2018). Soil biological indicator as an index for assessing the mango orchard sustainability. In: National Conference on Strategies & Challenges on Doubling Farmers' Income through Horticultural Technologies in Subtropics, held during June21-22, 2018 at CISH, Lucknow, pp. 133-134.

Babu, N., Adak, T., Arvind, K. and Singh, V. K. (2021). Scientific analysis of drought assessment and its management strategies for food security. Current Advances in Agricultural Sciences (An International Journal) 13(1): 1-6.

Dalal, R. P. S., Beniwal B. S., Rana G. S. and Dahiya S. S. (2017). Impact of split application of fertilizer at various growth stages on Kinnow productivity under semi-arid irrigated ecosystem. Indian J. Hortic. 74(2): 198-203.

Damodaran, T., Rajan, S., Ram Gopal, Yadav, A., Yadav, K., P. K. Shukla, Muthukumar, M., Kumari, N., Ahmad, I., Jha, S. K., Nayak, D. and Mishra, V. K. (2019). Successful community-based management of banana wilt caused by *Fusarium oxysporum* f. sp. cubense Tropical race-4 through ICAR-FUSICONT. Journal of Applied Horticulture, 21(1): 37-41.

Gundappa, Adak, T. and Shukla, P. K. (2018). Population dynamics of mango mealybug (Drosicha mangiferae) (Margorididae: Hemiptera) and its relation with weather parameters in subtropical climatic conditions. Indian Journal of Agricultural Sciences 88 (6): 865–870.

Gundappa, Adak T and Shukla PK. (2016a). Seasonal dynamics of mango hoppers and their management under subtropics. GERF Bull Biosc. 7(1):6-9.

Gundappa, Tarun Adak and PK Shukla. (2016b). Seasonal incidence and weather based forecasting model for mango leaf webber under subtropical conditions. Current Advances in Agricultural Sciences 8(1): 96-99.

Gundappa, Tarun Adak and PK Shukla. (2016c). Humid Thermal Ratio as a tool to assess mango thrips dynamics under subtropical climatic condition. J Environ Biol. 37:1239-1245.

Jayanthi Mala B. R., Kamala Jayanthi, P. D., Subramoniam, A. and Rekha, A. (2019). Management of Eulophid Seed Borer, Anselmella kerrichi (Narayanan *et al.*) (Hymenoptera : Chalcidoidea : Eulophidae) on Jamun. J. Hortl. Sci. 14(2): 166-168.

Khan, R. M., I. Ahmad, H. K. Kumar, A. Singh and P. K. Shukla (2019). Infestation of root-knot nematode, Meloidogyne enterolobii in newly established/old orchards and nurseries in Madhya Pradesh, Rajasthan and Uttar Pradesh. Annals of Plant Protection Sciences, 27(1): 170-171.

Kumar, A., Adak, T., Muralidhara B M and Singh, V. K. (2016). Enhancing farmer's income through strawberry cultivation under subtropical climate – a new initiative. GERF Bull Biosc. 7(2):12-16.

Kumar, A., Singh, N., Misra, K. K. and Vishal, N. (2017). Effect of foliar spray of calcium chloride and boric acid on shelf-life of guava. Indian J. Hortic. 74 (4):586-590.

Malhotra, S. K. 2017. Horticultural crops and climate change: A review. Indian Journal of Agricultural Sciences 87 (1): 12-22.

Manjunath, B. L., Laxman, R. H., Upreti, K. K. and Raghupathi, H. B. (2017). Partial root zone drying irrigation in papaya (*Carica papaya* L.) for enhanced water use efficiency under limited water situations. J. Hortl. Sci. 12(2): 143-149.

Nidhi Kumari, P. K. Shukla, Haripal Singh, Tahseen Fatima and Anju Bajpai (2022). Development of species specific PCR based detection assay for Ceratocystis fimbriata, mango wilt pathogen Indian Phytopathology 75: 215-220. https://doi. org/10. 1007/s42360-021-00438-9.

Nidhi Kumari, Thukkaram Damodaran, Israr Ahmad, Shailendra Rajan, P. K. Shukla, Muthukumar Manoharan, Amar Kant Kushwaha, Haripal Singh, Ram Gopal, Sangeeta Kumari, Kavita Yadav, Popy Bora, Sunil Kumar Jha (2023). Distribution and diversity of Fusarium oxysporum f. sp. cubense TR4 causing banana wilt in Subtropics of India and comparative analysis of TR4 specifc molecular detection methods. Journal of Plant Biochemistry and Biotechnology https://doi. org/10. 1007/s13562-023-00842-4.

Pal, R. and Ghosh S. N. (2018). Influence of micronutrients on growth dynamics, fruit yield and quality of Arka Neelamani grape. Indian J. Hortic. 75 (1):149-152.

Rashmi, I., Meena, H. R., Kala, S. and Kumawat, A. (2020). Micronutrient imbalance of sapota orchard in table lands of semi arid region of Rajasthan, India. Indian J. Hortic. 77(3): 553-555.

Rozante, J. R., Gutierrez, E. R., da Silva Dias, P. L., de Almeida Fernandes. A., Alvim. D. S. and Silva, V. M. (2020). Development of an index for frost prediction: Technique and validation. Meteorol Appl. 27: e1807.

Sahu, A., Kishore, K., Dash, S. N., Sahoo, S. C., Nayak, R. K. and Barik, S. (2022). Calcium nutrition influencing yield and fruit quality of dragon fruit. Indian J. Hortic. 79(3): 317-322.

Sharma, R., Jain, P. K. and Sharma, T. R. (2016). Improvement in productivities and profitability in high density orchard of mango (*Mangifera indica* L) cv. Amrapali through integrated nutrient. Economic Affairs 61(3): 533-538.

Shukla, P. K., A. K. Bhattacharjee and A. Dixit (2018). Field efficacy of difenoconazole against shoulder browning disease of mango (Mangifera indica L.) and its residue analysis for safety evaluation in mango fruit. Indian Phytopathology 71(1): 147-151.

Shukla, P. K., Adak, T., Misra, A. K. and Achal Singh (2016). Appraisal of shoulder browning disease of mango (Mangifera indica L.) in subtropical region of India. J. Mycology & Plant Pathology 46(1): 38-46.

Shukla, P. K., Adak, T. and Gundappa (2016). Dynamics of powdery mildew in mango and its evaluation using humid thermal index: an appraisal. In: Climate Change and Its Implications on Crop Production and Food Security. Editor Ratnesh Kumar Rao, P. K.

Sharma, B. Jirli, and M. Raghuraman Published by Mahima Research Foundation and Social Welfare, 194, Karaundi, Banaras Hindu University, Varanasi, UP, India pp. 351-355. (ISSBN 978-81-926935-4-5)

Shukla, P. K., Adak, T. and Gundappa (2016). Seasonal dynamics of powdery mildew of mango and its management under subtropics. GERF Bulletin of Biosciences 7(1): 21-25.

Shukla, P. K., Adak, T. and Gundappa (2017). Anthracnose disease dynamics of mango orchards in relation to humid thermal index under subtropical climatic condition. J. Agrometeorology 19(1): 56-61.

Shukla, P. K., Fatima, T. and Gundappa (2019). Efficacy of sterilized culture filtrate of Trichoderma harzianum Rifai against mango wilt pathogen, Ceratocystis fimbriata Ellis and Halst. Journal of Eco-friendly Agriculture, 14(2): 69-71.

Shukla, P. K., G. Baradevanal, S. Rajan and T. Fatima (2020). MaxEnt Prediction for potential risk of mango wilt caused by Ceratocystis fimbriata Ellis and Halst under different climate change scenario in India. Journal of Plant Pathology, https://doi. org/10. 1007/s42161-020-00502-9.

Shukla, P. K., Nidhi Kumari, Tahseen Fatima and Haripal Singh (2023). Evaluation of bio-control agents for the management of guava decline. Journal of Eco-friendly Agriculture 18(1): 193-199.

Shukla, P. K., S. Varma, T. Fatima, R. Mishra, A. K. Misra, A. Bajpai, Gundappa and M. Muthukumar (2018). First report on wilt disease of mango caused by Ceratocystis fimbriata in Uttar Pradesh. Indian Phytopathology 71(1): 135-142.

Shukla, P. K., T. Fatima and N. Kumari (2020). First report of Berkeleyomyces basicola causing mango root rot and decline in India. Plant Disease 105(4): 1214.

Shukla, S. K., Adak, T. and Singh, V. K. (2020). Evaluation of boron nutrition in enhancing productivity and quality of mango (*Mangifera indica* L.) cultivar Mallika under subtropical climatic conditions. Current Advances in Agricultural Sciences 12(1): 33-36.

Shukla, S. K. and Adak, T. 2020. Intercropping of turmeric for revenue generation from interspaces in mango orchards. Kahaar 7(1-2): 41-43.

Shukla P. K., Adak, T., Gundappa and Misra A. K. (2016). Weather based models in predicting powdery mildew of mango for Malihabad mango belt. Journal of Eco-friendly Agriculture, 11(2): 142-148.

Singh, A. K., Singh, S., Appa Rao, V. V., Hiwale, S. S. and Joshi, H. K. (2015). Long term effect of INM on aonla (*Emblica officinalis*) and soil quality under rainfed hot semi-arid environment. Indian Journal of Agricultural Sciences 84(5): 585-588.

Tehranifar, A. and Tabar, S. M. (2009). Foliar application of potassium and boron during pomegranate (*Punica granatum*) fruit development can improve fruit quality. Horticulture, Environment and Biotechnology 50(3): 191-196.

3

Canopy Management in Old and Unproductive Mango Trees as an Advisory for Growers Benefit

Sushil Kumar Shukla, Dushyant Mishra and G. Pandey

ICAR-Central Institute for Subtropical Horticulture, Lucknow, Uttar Pradesh

Introduction

Mango (*Mangiferaindica* L) known as "the king of fruits" and "Apple of the Tropics" is most popular fruit in the tropics and sub-tropics. It moved from India to different countries and has gained popularity around the world. It has long been associated with the Indian culture and religious functions. Among different states, Uttar Pradesh and Andhra Pradesh constitute 23.64 per cent and 22.99 per cent share in the total mango, respectively, in the country. In 2023, Mango production had a total export volume of 78.56 billion MT. India is the biggest mango producer with a total production volume of 25 million MT of mangoes during the year. The top 10 mango producers in 2023 include India (25 million MT), China (3.8 million MT), Indonesia (3.6 million MT), Pakistan (2.7 million MT), Mexico (2.4 million MT), Brazil (2.1 million tons), Malawi (1.7 million MT), Thailand (1.6 million MT), Bangladesh (1.5 million MT) and Vietnam (1.4 million MT). As regards world-wide mango exports, Mexico is the biggest mango exporter with total exports worth $576.36 million in 2023. It was noted that Mexico's mango exports increased by 5.09% from the previous year. The top 10 mango exporting countries in 2023 include the following:

S. No.	Name of the country	Value of mango exports (US $) in 2023	Share (%)	Increase/Decrease over previous year
1	Mexico	575.36 million	38.08%	+5.09%
2	Netherlands	381.72 million	25.27%	+ 15.53%
3	Brazil	315.69 million	20.90%	+ 52.62%
4	India	148.36 million	9.8%	-38.96%
5	United States	48.23 million	3.19%	- 4.63%
6	Germany	40.24 million	2.66%	+10.60%
7	Burkina Faso	27.63 million	1.83%	-21.94%
8	Portugal	27.22 million	1.80%	+26.60%
9	South Africa	17.18 million	1.14%	-11.38%
10	United Kingdom	12.82 million	0.85%	+48.20%

Source: Google https://www. tradeimex. in/blogs/top-10-Mango-exporter

Major concern in export of mangoes is that of poor quality standards which limit India's share in international market. Old and unproductive orchards with huge tree size and without proper canopy management is the major reason responsible for poor mango productivity and quality especially fruit size and chemical residues. Canopy management also helps in minimizing no of chemical sprays and thereby reducing the possibility of toxic residues in fruit pulp. Such dense and unproductive mango orchards constitute 35-40% of total area in mango growing belts. In mango belts like Malihabad, mango orchards have been almost converted into mango forest like plantations. As we know, fruit trees, invariably, after a certain period of successful production, have a tendency to lose their productivity and quality making the orchards unproductive and non-remunerative. Replacing the old and unproductive orchards through replanting is a long term strategy which requires many years to reach the desired level of productivity. Rejuvenation technology for such unproductive trees offers many benefits such as early and quality production from such trees through better light penetration and efficient utilization of available space and inputs. Keeping this in view, a mango rejuvenation technology was developed by ICAR CISH about 20 years back and it was promoted adopted throughout the country through state horticulture departments/National Horticulture Board/National/State Horticulture Mission. However, mortality of trees and primary branches as a result of heavy stem borer infestation has been the major constraint, besides its misuse through deliberate negligence in the after-care of pruned trees for clearing road-side mango plantations with a view to converting land use. So, the need was felt to develop a new technology to address these issues. In light of the above experiences, efforts were made to develop a new technology which could address the relevant issues.

After a study and demonstration trails at farmers' field, a safe and fruitful technology has been developed for rejuvenating old and unproductive mango orchards in such a way that farmers don't lose the fruit yield even in the process of rejuvenation and the problem of tree mortality could also be solved. This is known as refined mango rejuvenation technology as it has bright prospects to address the farmers' issues (Fig. 1).

Fig. 1: Profuse flowering and Fruiting in 60 year old rejuvenated mango trees at farmer's field

A Thoughtful of Mango Rejuvenation Technologies from Different Parts of India and Abroad

Mango trees generally begin commercial production in three to four years after planting and continue to produce progressively increasing yields as canopies enlarge until shading by adjacent trees forces the growth upwards and out of reach of harvests. Lower branches supporting previous year's productive stems become unproductive due to shading by branches located higher. Flowering and fruit production in mango occurs on stem terminals; therefore, as canopy size increases, production moves to the top as competition for available light continues. Like other fruit trees, mango trees fail to sustain their productivity and economic viability after a certain age. Sometimes it may be natural or a number of factors are responsible for deterioration in the productivity. It has been observed that after about 45-50 years of planting, the branches of one tree start intermingling with those of the others. As a result, the orchards become dense and are devoid of sunlight affecting the photosynthesis process of the leaves. Such a condition leads to poor and sickly growth of shoots, reducing their potential to bear normal fruit load. Whatever new shoots emerge, they become weak and are unsuitable for flowering and fruiting. Dominance of such orchards is one of the major reasons for low mango productivity. In our country, about 35-40 per cent mango orchards are old and senile. By the age the trees reach 45 years, they have grown tall with less canopy area and that too only on the top. Such trees harbour insect-pests and diseases and it becomes difficult to control them easily. Pruning of trees in old orchards helps develop new shoots which grow into a desired canopy, increases light interception by the foliage. It also helps in an effective management of insect pests and diseases, restoring their vigour and productivity.

Importance of Rejuvenation in Mango Trees

In order to get rid of old unproductive orchards existed in abundance, large scale uprooting and replanting is a long term and expensive strategy. Such trees can be made productive and profitable through rejuvenation technique as it restricts the further deterioration of the health of plant, extending its life and improving the yield of quality of produce within minimum time as compared to establishment of new orchard which attains the commercial bearing stage after 8-10 years (Kalloo, *et al.*, 2005; Lal and Mishra., 2008, Singh, 2007, Hiwale, 2009, Pandey *et al.*, 2013; Shukla *et al.*, 2016). Rejuvenation technology is a unique integration of invasive exhaustive operations and skillful management for transforming the trees from unproductive to productive ones by restoration of youthful vigor (Fig. 2). It deals with many issues like restoration vigor, enhancement of quantum and quality production, sustainability and economic feasibility, etc. All these aspects have been proved with sound experimentation, implementation and success stories (Singh and Mishra, 2007; Pandey *et al.,* 2013). Reiterative pruning and canopy rebuilding for rejuvenation of perennial fruit trees is medium term strategy for making the senile trees productive in a sustainable way within a time gap of three years for mango (Lal *et al.*, 2000; Kumar, 2008).

Insights of Physiology of Rejuvenation

Perennial fruit trees build their woody architecture with some building blocks, i. e. modular units, known as metamer or phytomer (White, 1979; Room *et al.*, 1994; Hare Krishna, 2012), which consists of meristems, initially an apex. When a meristem faces periods of dormancy during low temperature or pruning, it appears as bud, whenever active, it produces a growing point with sequence of meristematic tissues simultaneously called a sprout, i. e. extension unit or growth unit (Halle *et al.*, 1978; Bell, 1991). The driving mechanism behind the pattern appears to be auxin production when the new shoot and leaves developed and are exposed to light (Sterck, 2005). Other features of trees are to produce more xylems relative to phloem, accumulate xylem over time and may thus produce thick woody branches and stems. The studies have also suggested that attaining reproductive phase and reproduction processes compete with vegetative functions, depleting resources necessary for maintenance and growth (Fakhri *et al.*, 1987). The concept of source and sink helps to explain allocation patterns at both the physiological and evolutionary levels. The allocation to vegetative structures has compounding interest for plant growth, as plant life cycles begin with pure vegetative growth (Abrahamson, 1979). Photosynthesis either directly or indirectly provides the building blocks for all biomass production and growth but the plant growth and productivity are determined more by the way in which the product of photosynthesis are allocated and utilized during plant growth and development.

The decline in productivity of orchards is largely due to poor photosynthetic efficiency besides several other factors including age of orchard, dense and intermingling branches, neglect and poor management leading to production of poor quality fruits or unproductive stage. This requires formative or corrective pruning to reinforce the tree to solve the bearing problem of old orchards, i. e. making them productive again. This is what is done by rejuvenation technology. After severe pruning, the left over thick branches of the tree are the primary source of reserve carbohydrates in the absence of shoots and leaves (Hare Krishna, 2012). The appearance of sprout, emergence of shoots and young leaves from the pruned branches of the old tree depend on imported carbohydrate from the sieve tubes of the phloem and other tissues to provide metabolism and carbon skeletons for metabolic inter conversions and bio-synthetic process. Vegetative re-growth as sprout emergence is a sink for resources, and so allocation to it must be under ecological controls similar to those that govern sexual reproductive allocation. There is existence of meristematic tissues in the left over trunk portion after heavy pruning, which are capable of active cell division and differentiation into specialized tissues as shoots and leaves at initial stage importing stored carbohydrates (Singh, 2003;

Hare Krishna, 2012). New shoots emerge after pruning but they don't possess flower buds for one to two years. The developing new shoots of current season continue growth taking enough time to mature to for phase change. It experiences dormancy during exposure to winter or low temperature for one or two seasons thereby changing to floriferous condition as its vegetative meristem has differentiated into inflorescence (Singh, 2003; Shukla *et al.*, 2016).

It is necessary to develop a framework strong enough to support the load of fruit yield. The consequences of severe pruning, the method and time gap of rejuvenation coupled with cultural practices are required to support the reinvigoration process (Kumar, 2012). Thinning of emerged shoots after six months of growth and selection of primary scaffolds are done to properly train them outward to grow fast and outward. Scaffolds should be headed during the dormant season of first two years to promote lateral branching to be stiffened and strengthened as well as for maturing the scaffolds for shaping desired framework (Kumar, 2015). Floral induction in some perennial fruit crops like mango, litchi, guava, citrus, etc occur in response to cool temperatures perceived by mature leaves largely regulated on physiologically mature shoots (Menzel and Simpson, 1995; Menzel *et al.*, 2000). Periodic monitoring of trees after pruning operation, emergence of re-growth and its maintenance avoid complexities and disappointments due to undesired framework development or chances of mortality (Kumar, 2008).

Fig. 2: Demonstration of rejuvenation technology as an advisory for farmers benefit

Insights of Old CISH Mango Rejuvenation Technology

The CISH mango rejuvenation technology (Kalloo, *et al.*, 2005; Shukla *et al.*, 2016) aimed at pruning of undesired branches for inducing development of umbrella like open canopy of healthy shoots. Open canopy ensures better light penetration, improves photosynthesis and flowering capacity, thereby fruiting potential of shoots. Pruned trees attain canopy of healthy shoots in two years and from third year onwards they start bearing fruits. Following steps are involved in the process of mango rejuvenation:

1. Identification of unproductive mango orchards by yield records of last five years
2. Heading back of 3-5 wide angled primary branches leaving about 75 cm from their base during December at a height of 4-5 meters to develop umbrella like tree canopy and thinning out remaining branches from their base.
3. Heading back/pruning should be done with power or petrol operated chain saw to ensure sharp and smooth cuts. A fungicidal paste prepared by mixing 1 kg copper oxy chloride, 250 ml castor oil and water to desired consistency, is applied on the cut surfaces immediately after pruning to check microbial infection.
4. Fertilizer application, viz., 2.5 kg Urea, 3.0 kg Single Super Phosphate & 1.5 kg Muriate of Potash along with 50 kg FYM should be done in February followed by light irrigation. Need based mulching and irrigation should be done at times.
5. Thinning of excessive shoots during April and June to allow only 6-8 healthy shoots per branch for ideal canopy development.
6. Monitoring and management of insect pests like stem borer, leaf eating caterpillars and diseases like anthracnose by need based spray of recommended insecticides/fungicides.
7. After three years of heading back, the trees become rejuvenated with development of healthy and productive canopy and start bearing fruits.
8. Orchardist can get additional income from sale of pruned wood and intercropping of short duration crops such as vegetables and flowers in open space of orchard during the initial years after hard pruning.
9. Fruit yield is enhanced up to 4-5 times and quality is also improved with bigger fruit size. Higher yield is due to better harvesting of solar radiation by the tree canopy.

Growth of emerging shoots on pruned branches was influenced by severity of pruning (Lal *et al.,* 2000). Length and girth of emerging shoots were more in first, second and third order pruned trees than in fourth and fifth order treatments, whereas it was minimum in the control. Fruit yield was higher in fourth and fifth order pruned trees than the control. No fruit yield was obtained in first and second order pruned trees during first and second year of fruiting, but yield was obtained in an increasing trend in subsequent years. Cumulative fruit yield of 6 years indicated that fourth order pruned trees gave the highest yield, whereas it was lowest in the control. However, further studies continued by Lal and Mishra (2008) revealed that after twelve years of pruning, highest increase in tree height was recorded in first order pruned trees. Highest shoot length was recorded in fourth order pruned trees, while shoot girth was highest in first order pruned trees. Canopy spread was the highest in control. There was loss of fruiting for initial two years in severe pruning treatments (first and second order pruned trees), while from third year onwards; there was a gradual increase in fruit yield in these treatments. In moderate and light pruning treatments (third to fifth order pruned trees), there was continuous fruiting from first year onwards with an increasing trend for initial 3-4 years, there after inconsistent yields were obtained.

The mango rejuvenation is rather critical and somewhat difficult and under high rainfall conditions it is not recommended, since the tree may not recover after pruning and thus it requires intensive care and attention. The technique is common in sub humid and semi-arid areas. However, reports suggest that in warm humid tropical areas this technique is not widely used because of subsequent loss of fruit production and excessive re-growth (3 to 4 vegetative flushes per year) that occurs (Crane *et al.*, 2009). Rejuvenation pruning studies conducted in old non-bearing trees of mango cv. Alphonso revealed that heading back operation along with paclobutrazol @7.5g a.i. /tree and canopy rebuilding has been found beneficial for earliness of flowering and getting higher fruit yield (Mistry and Patel, 2009; Kumar and Kavino, 2012). Banik and Sen (2002) successfully rejuvenated the unproductive trees of mango cv Fazli aged 30-50 years with the rejuvenation pruning comprising leader intact, removal of first order branch and pruning of secondary branches and application of 1.0 kg each of N and K_2O and 0.5 kg P_2O_5/plant/year along with 3% spray of urea in September, resulting in maximum growth and yield (388.42 kg/plant) of older trees above 50 years of age. Depending upon the extent of pruning, there are four model approaches that have been in practice at various locations to take up rejuvenation in mango (Hackett, 1985; Shukla *et al.,* 2016):

Model-1

In this model, tall central trunks are cut back to about 3-4 m height from the ground level. The actual site to cut back to should be at a point where there are side branches. The operation is performed using a chain saw to give a smooth cut. It is a partial rejuvenation where one half of the tree (all around) is cut back and the remaining trunks and leaves will help protect the stump from sunburn (Crane *et al.*, 2009; Balamohan *et al.*, 2011). The freshly exposed trunks and branches to the sun should be painted with white water based paint diluted three or four times. This is to prevent sunburn, which could attract borer onto the damaged bark. In short time, numerous shoots will develop. Most vigorous shoots spaced evenly around the stump and if possible at different heights, are selected, removing all the unwanted shoots. This process should be repeated as often as necessary until the selected shoots begin to dominate and take over. Then the rest of the tree can be cut back and the operations may be repeated. By this method, half the yield will be realized throughout the process till it regains the vigour. Stassen (1999) compared the different systems of pruning with un-pruned controls under high density orchards of mango and concluded that about closed vase system (informal pyramid) performed the best.

Model-2

Old, unthrifty trees can often be rejuvenated by a moderate to severe pruning. This is the form of skeletonizing the tree, i. e. cutting back the branches of the tree till only basic frame is left. Moderate skeletonizing would entail cutting back the large branches to healthy wood, while at the same time maintaining the basic structure of the tree. Particularly, large unthrifty trees would benefit from more severe skeletonizing, where all the main branches are cut back to the main trunk only leaving about a meter or so of branch. The whole trunk and remaining branches should be painted with diluted paint to prevent sunburn and borer attack. In both these cases, there will be proliferation of sucker growth (Balamohan *et al.*, 2011)

Model-3

In this case, the extent of pruning is decided on the basis of first, second, third and fourth order branches. It was found that for rejuvenation, the pruning treatment with paclobutrazol application recorded significantly higher and quality fruit yield in fourth and fifth order pruned trees than the control and succeeded in restoring the health, vigor and sustainable fruit yield in declining mango orchard (Burondkar *et al.*, 2000; Lal *et al.*, 2000, Shinde *et al.*, 2002). Reddy and Kurian (2011) recorded significantly higher and almost double mean cumulative yield with third order pruning in mango cv Alphonso when compared with un-pruned control. At IIHR, Bengaluru, pruning third order branches 30 cm from point of origin recorded the maximum yield in Alphonso (86.3kg/tree) (Fig. 3). Trial conducted at CISH revealed that pruning second order branches recorded the maximum pooled fruit yield of 57.89 kg/tree from cv. Dashehari (Lal and Mishra, 2008; Balamohan *et al.,* 2011)

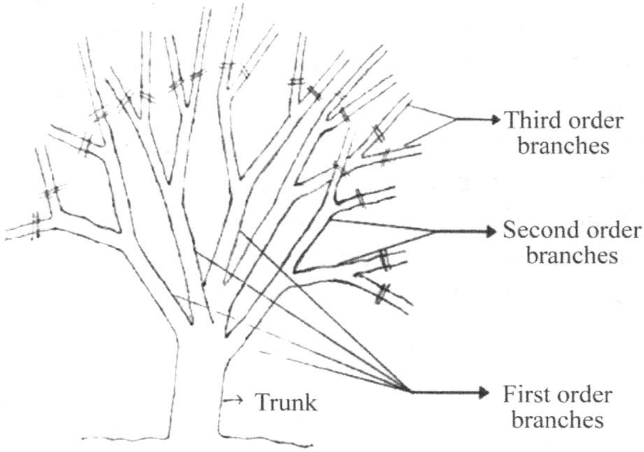

Fig. 3: Tertiary branch pruning retaining 30 cm from base as advocated by IIHR (Diagram Courtesy-Balamohan *et al.*, 2011)

Model-4

In this model, reiterative pruning is done in a phased manner up to the height of 3-4 m from the ground level and later canopy rebuilding in open umbrella shape with intensive care and management transforming successfully from unproductive to productive phase with enhanced quantum of quality fruit production within a gap of three years onwards (Kalloo *et al.*, 2005; Kumar and Kavino, 2012). Studies were conducted at CISH, Lucknow to rejuvenate 40-50 year old mango trees in the farmer's field. In this process, main branches were pruned at a height of 5 m from the ground level during December. Pruned surfaces were smeared with fungicidal paste immediately after pruning to check microbial infections. Pruned trees were kept under intensive care and management. Cultural practices and plant protection measures were adopted properly. Profuse shoots emerged and outward growing 8-10 shoots were retained per branch, removing the undesired ones. Since pruned trees started fruiting from third year, the growers ended up with loss by missing crops for two years, hence technology was refined to undertake the pruning work in alternate rows in the orchards. With alternate row pruning, availability of light to unpruned trees in two adjacent rows was greatly improved and their fruiting increased by 2-3 times. Thus, enhanced production from unpruned trees compensated the loss to some extent (Balamohan *et al.*, 2011).

Constraints and Feedback on Old Rejuvenation Technologies

In previous mango rejuvenation technologies, it was advocated to head back all the primary branches of the tree in one go during winter months (December-January) and ensure proper after-care of trees to ensure proper success. The major constraint in this technology was the problem of mortality of trees/primary branches due to heavy infestation of stem borer which farmers were often not able to manage. There were cases where the technology was misused/abused and it was used for clearing road side mango plantations for changing the land use by deliberate negligence in the after care of trees after rejuvenation pruning. A study on the impact of old CISH Rejuvenation technology from 12 different sites revealed average survival only to the tune of 45.7% after rejuvenation pruning (Table 1).

Table 1: Site-wise status of rejuvenated mango trees

Sl. No.	Villages	No of farmers adopted	No of plants rejuvenated	No of plants survived	Percent of trees survived
1	Kakori	4	66	42	63. 63
2	PahiyaAzampur	3	15	nil	nil
3	Sahilamau	4	100	14	14.00
4	Golakuan	5	99	47	47.47
5	DurgaGanj	1	18	02	11.11
6	Mandauli	3	87	60	68.96
7	Dalel Nagar	1	21	0	0
8.	SIMA, Rehmankhera	1	9	0	0
9	Khalispur	2	30	1	3.22
10	Saidpur	2	108	98	90.74
11.	Banswara	1	80	70	87.50
12.	Sangareddy	1	340	339	99.70
	Total	**28**	**996**	**655**	**45.70**

The study of such cases revealed lack of proper after-care of orchards. Particularly, after cutting of primary branches for induction of new shoots care is a must. The farmers with relatively small land holdings who could manage the infestation of stem borer with the recommended measures were successful while, the farmers, who neglected the orchards after pruning operation, had the problem of major tree mortality. Many times, it is not possible to manage the stem borer infestation on large scale in bigger orchards. Besides, when we go for such rejuvenation, farmers either completely lose the crop in the initial 2 to 3 years due to shoot growth and canopy development or get little yields from third year onwards. Keeping this in view, efforts were made to develop a new technology for enhanced success and profitability.

Insights of Refined Mango Rejuvenation Technology

The new refined mango rejuvenation technology aims at thinning out central leader and other undesired branches and heading back of selected 3 to 5 primary branches well distributed in different directions in a sequential manner over a period of 3 to 5 years and starting from top of the tree canopy. In the first year, two primary branches situated in east and west directions are headed back in the month of December-January followed by heading back of two branches of north and south direction in the following year while the remaining branch is headed back in third year. This ensues regular income to the farmers from the remaining branches. The detailed steps of the technology are given below (Shukla *et al.*, 2021).

Selection of Orchard

The exact reasons for the decline or poor productivity of orchards should be determined and corrected, if possible. Otherwise, response to rejuvenation will be temporary. If the trunk and basic scaffold limbs of trees are not structurally sound due to disease, heart rot and cold injuries, the rejuvenation may not be justified. A limited root system due to high water table, a hard pan or poor subsoil can result in early tree decline with trees being more readily stressed by drought. When this occurs, the top becomes out of balance with the more limited root system with consequent insufficient uptake of water and nutrients by the roots. Pruning such trees will temporarily restore them to a more favorable top to root ratio, but not alleviate the long term problem. It has been observed that orchards of mango planted at 10×10m, after attaining the age of 30-40 years turn dense with compact top canopy cover and long barren (leafless) branches and bearing confined only to the top periphery. It is often uneconomical to maintain old orchards after 30 to 40 years of age due to production of poor quality fruits on high tops which poses problem in harvesting and other horticultural operations, increased incidence of diseases, insect pests and physiological disorders, alternate bearing habit. Such orchards need to be rejuvenated for further continued production of quality fruits (Kumar, 2015). Sometimes, the condition of orchard is so poor that it is not at all profitable to rejuvenate. It is advisable to go for new plantation in such cases instead of rejuvenation or the efforts should be made for satisfactory growth maintenance by other required cultural and management practices (Singh, 2007).

The Pruning Operation

First of all, 3-5 primary branches well distributed in all directions are selected and marked on the tree for heading back operation in a sequential manner over a period of 3 to 5 years. Secondly, the centrally located upright growing branch of the tree, if present, is to be thinned out, completely. This central leader will be removed from the point of its origin without leaving any portion of it. Thus, efforts are made to have a canopy with open centre which facilitates light penetration inside the canopy. All other branches growing very low in height and interfering with the horticultural operations or intermingling with the selected branches will also be removed from the point of their origin (Plate 1).

In the first year, after thinning out central leader, two primary branches of east and west direction and located on the top are headed back leaving a stump of 1 to 1. 5m for regeneration of secondary branches (Plate 1 and 2). This heading back is done at a height of 3.5 to 4.5 m from the ground level depending upon the tree architecture and availability of branches. In case the operation is to be completed over a period of five years, only one primary branch growing in east direction will be headed back. Remaining all the selected branches will be maintained for fruiting. In the second year, two branches situated in north and south direction will be headed back (Plate 3 and 4) in the similar fashion while only one branch situated in west direction will be headed back where the operation is to be completed over a period of five years. In the third year, remaining one branch will be headed back (Plate 4 and 5) while the pruning operation will continue every year over a period of five years if you wish not to lose fruit yield from the remaining branches in the initial years.

Time of Pruning

The ideal time for performing the pruning operation in mango is from December to mid January in north India as delayed beheading of branches leads to oozing of sap and gum due to high temperature while there are chances of infection/insect pest attack during rainy season which warrants more number of chemical sprays. Pruning can also be done just after rainy season i. e. during September-October in induction of shoots at an early date but it required proper care of newly emerged shoots from both insect-pests/diseases and frost during winter.

Plate 1: Original trees before pruning

Plate 2 : Tree after heading back two opposite branches

Plate 3 : Fruiting from remaining two branches

Plate 4: Tree after heading back four branches and fruiting from remaining branch

Plate 5: Trees after completion of pruning operation

Protection of Cut Wounds

In north India, December-January months are the safest months from the view point of minimum microbial infections due to prevalence of low temperature. However, it is recommended to apply Bordeaux Paste on the cut wounds immediately after pruning. Bordeaux paste can be prepared dissolving 100 g each of powdered copper sulphate and quick lime, each in 500 ml of water separately. When both the solutions are ready, they are mixed together to make 1 kg of paste. The cut portions will be smeared with this paste. It facilitates the healing of cut portions without any further infection. Alternatively, fungicidal paste can also be prepared by mixing 1 kg of copper oxychloride/copper hydroxide with 250 ml castor oil/neem oil and water to desired consistency.

Orchard Management and after Care

Nutrition: There should be integrated nutrient management for maintenance of soil fertility and plant nutrient supply to an optimum level for sustaining the desired growth and productivity level. It is necessary that fertilizer applications are made using both organic and inorganic forms in pruned orchard with the sole purpose of initiating new shoots with vigor at an early date. At the time of pruning, trees should be fed with optimum fertilizer doses, followed by irrigation near the root zone, just like the commercial bearing orchards. The mango trees are well fed with recommended doses of manures and fertilizers and other after-care well before the cutting operation. A dose of 100 kg well rotten FYM is applied just during rainy season while a dose of 2.0 kg of urea, 3 kg of single super phosphate (SSP) and 1 kg of Muriate of Potash (MOP) is applied for each pruned tree just after rainy season latest by September when sufficient soil moisture is available. A second dose of 500 g each of urea and Muriate of Potash (MOP) will be applied after fruit set at proper moisture level. Adjustment in the doses of nitrogenous and phosphatic fertilizers should be made if Di-ammonium phosphate (DAP) is used in place single super phosphate (SSP). Besides this, need based application of micronutrients like zinc, copper, boron should be ensured for quality mango production from the first year itself. Foliar nutrition by spray of fertilizers like NPK (19:19:19) in month of September is also beneficial for both emergence of new shoots after pruning and also for fruiting on remaining branches. Manures and fertilizers are applied in the tree basins around the trees leaving 1.0 m distance from the trunk. The fertilizers should be mixed properly in the top soil up to the depth of about 15 to 20 cm. Basins and irrigation channels should be prepared during January after pruning operation. Weeding in the tree basins should do from time to time to keep the orchard weed free before application of fertilizers. Chemical fertilizers can also be applied in 25 cm wide and 20 cm deep trench made 2.5 m away from the trunk around the tree.

Irrigation and Mulching

Optimum soil moisture should be done by regular irrigations during winter to avoid frost damage and summer for proper fruit development and emergence of new shoots on the pruned branch/branches. As sufficient numbers of branches are left for fruiting in the first year, care should be taken to avoid irrigation just before flowering and give regular irrigations at 15-20 days interval after fruit set till maturity. Mulching is done in the tree basins to maintain soil moisture during hot summers and also cut down the number of irrigations.

Thinning of Shoots

There is profuse emergence of new shoots on the pruned branch (branches) with the rise in temperature during March. If all the shoots are allowed to grow, there would be undesirable competition among the shoots for space, light, nutrition and growth. This would also lead to dense and bushy shoot growth on the pruned branches. The pruning brings invigorating effect and subsequent emergence of numerous shoots on the cut branches requires proper thinning. Thinning of shoots and canopy should be skillfully done by just removing the extra, overlapping down growing, weak shoots after a certain gap period. Thinning of excessive shoots arising from cut stumps is started from April onwards to allow only 6 to 8 healthy shoots per branch for ideal canopy development. Best time for thinning of undesired shoots is June and September. Smearing cut portions with the fungicidal paste would be appropriate after thinning of shoots.

Application of Paclobutrazol

Application of paclobutrazol @ 16 g/tree in the 20 cm deep trenches made 1.5 to 2 m away from the tree trunk in the month of September helps in the fruiting of remaining branches on regular basis. The dose is reduced to half, i. e. 8 g/tree in the following year to ensure fruiting on the mature shoots as well as on the remaining branches. In the third year, no application of paclobutrazol is required. Thus, application of paclobutrazol in a cyclic manner coupled with proper nutrient management ensures regular flowering and fruiting from the pruned trees in the following years.

Annual Maintenance Pruning

After development of an umbrella type canopy over a period of 3-5 years, annual maintenance pruning is required to thin out the undesired branches and open up the centre for proper light penetration and contain the height and canopy size of the tree. In this pruning, dried, dead and criss-cross branches are removed besides thinning out a few branches from the centre of the tree canopy.

Survival and Growth Parameters

Cent per cent survival of trees and primary branches was recorded in the technology. Unpruned mango trees had the average height of 9.72 m, trunk girth of 1. 63 m, canopy spread in east-west direction 10.2 m and canopy spread in north south direction 11.3 m in Dec 2015. These parameters, viz., tree height, trunk girth, canopy spread in east-west and north-south directions rose to 10.9 m, 1.93 m, 16.1 m and 15.38 m, respectively. In case of rejuvenated trees, tree height varied from 7.25 m to 8.28 m while average canopy spread in east-west and north-south directions was around 8 m in 2023.

Fruit Yield

The study over a period of 7 years revealed highest mean cumulative fruit yield of 52.34 kg/tree in case of T-5 where the pruning was completed in three years, while cumulative mean fruit yield of 50.71kg/tree was obtained in case of trees where rejuvenation operation was completed in five years by heading back one branch/year. Varying fruit yield from 40 to 150 kg was obtained from the remaining branches after pruning operation in the process of rejuvenating mango trees (Anonymous, 2018, Shukla, *et al.*, 2021).

References

Abrahamson, 1979. Patterns of resource allocation in wildflower populations of fields and woods. American Journal of Botany 66: 71-79.

Anomymous 2018. Mango rejuvenation technology refined. CISH News Jul-Dec, No. 2: pp 1-2.

Balamohan, T, Mekala, P., Rajadurai, S., Priyadharshini, G. Prakash, K., Soorianathasundaram, K and Kumar, N. 2011. Canopy management in mango. Technical Bulletin, TNAU, Coimbatore: 28p.

Banik, B. C. and Sen, S. K. 2002. Effect of pruning and fertilization on growth and yield of mango cv. Fazli. The Orissa Journal of Horticulture 30(1):13-18.

Bell, A. D. 1991. Plant Form. Oxford University Press, New York.

Burondkar, M. M., Gunjate, R. T., Magdum, M. B. and Govekar, M. A. 2000. Rejuvenation of old and overcrowded Alphonso mango orchard with pruning and use of paclobutrazol. Acta Hort. 509:681-686.

Crane, J. H., Garcia, L. Salazor, Lin T. S., Pinto A. C. de Queiroz and Shu Z. H. 2009. Crop Production Management. (In) The Mango: Botany, Production and Uses, 2ndedn, pp 432-83. Richard E Litz (ed). CABI Publication, UK.

Fakhri, A., Bazzaz, Nona, R., Chiariello, Phyllis D. Coley, Louis F. Pitelka. 1987. How plants cope: Plant Physiological Ecology. Bioscience 37(1):58-67.

Hackett, W. P. 1985. Juvenility, maturation and rejuvenation in woody plants. Horticultural Reviews 7:109-155.

Halle, F., Oldeman R. A. A. and Tomlinson, P. B. 1978. Tropical Trees and Forests. An Architectural Analysis. Springer Verlag, Berlin-Hiedelberg.

Hare Krishna. 2012. Physiology of Fruit Production. Stadium Press (India) Pvt Ltd, p 101.

Hiwale, S. S. 2009. Orchard Rejuvenation. (In) The Pomegranate. New India Publishing Agency, New Delhi, pp 67-72.

Kalloo, G., Reddy, B. M. C., Singh, G. and Lal, B. 2005. Rejuvenation of old and senile orchards. Extension Bulletin-25, CISH, Lucknow, 40P.

Kumar, R. 2012. Adoption of improved production technologies by litchi growers for quality production. Journal of Community Mobilization and Sustainable Development 7(2): 210-216.

Kumar, Rajesh. 2008. Rejuvenation of old senile litchi orchard. Tech. Bull. No. 5. NRC for Litchi, Muzaffarpur, pp1-8.

Kumar, Rajesh. 2015. Rejuvenating old, senile orchards of tropical and subtropical fruits for enhanced production and improved quality: A Review. Indian Journal of Agricultural Sciences, 85(3):295-313.

Lal, B., Rajput, M. S., Rajan, S., Rathore, D. S. 2000. Effect of pruning on rejuvenation of old mango trees. Indian J. Hort. 57 (3): 240-242.

Lal, B. and Mishra, D. 2008. Studies on pruning in mango for rejuvenation. Indian J. Hort. 65(4): 405-408.

Menzel, C. M. and Simpson, D. R. 1995. Nutritional studies on lychee trees in subtropical Australia. Acta Hort. 273:581-585.

Mistry, P. M. and Patel, B. N. 2009. Impact of heading back plus paclobutrazol on rejuvenation of old and overcrowded Alphonso orchards. Indian J. Hort. 66(4): 520-521.

Pandey, D., Pandey, A. and Yadav, S. 2013. Horticultural interventions for increasing productivity of aonla and bael. (In) Recent Advances in Subtropical Fruit Crop Production, pp 35-39. Singh V. K. and Ravishankar, H. (Eds). CISH, Lucknow.

Reddy, Y. T. N. and Kurian, R. M. 2011. Studies on rejuvenation of old, unproductive Alphonso mango trees in orchards. J. Hortl Sci. 6(2):145-147.

Room, P. M., Maillette, L. and Hanan, J. S. 1994. Modular and metamer dynamics and virtual plants. Advances in Ecological Research 25:105-157.

Shinde, A. K., Waghmare, G. M., Godse, S. K. and Patil, B. P. 2002. Pruning for rejuvenation of overcrowded old Alphonso mango (Mangifera indica L) gardens in Konkan. Indian Journal of Agricultural Sciences 72(2):90-92.

Shukla, S. K., Mishra, D., Pandey, G. and Rajan, S. 2021 Rejuvenating old and unproductive mango orchards for enhanced farmers' income. Indian Hort., 66(4): 32-34.

Shukla, S. K., Pandey, G, BarsatiLal and Dushyant Mishra (2016) Rejuvenation of Fruit Plants as a Means of Improving Productivity and Profitability. (In) Doubling Farmers Income through Horticulture (Eds KL Chadha, SK Singh, P. Kalia, WS Dhillon, TK Behera, Jai Prakash). Daya Publishing House ((A Division of Astral International Pvt Ltd, New Delhi): pp 227-238.

Singh, A. 2003. Fruit Physiology and Production, pp 488. Kalyani Publishers, New Delhi.

Singh, B. 2007. Rejuvenation of old orchard. (In) Horticulture at a Glance, pp43-45. Kalyani Publishers.

Singh, G. and Mishra, R. 2007. Aonla Rejuvenation. PFDC, CISH, Lucknow, 16P.

Stassen, P. J. C., Grove, H. G. and Davie, S. J. 1999. Tree shaping strategies for higher density mango orchards. J. Appl. Hort. 1(1):1-4.

Sterck, Frank. 2005. Woody Tree Architecture. (In) Plant Architecture and its Manipulation, Annual Plant Reviews, pp209-237. Turnbell Colin G. N. (Ed) Blackwell Publishing, CRC Press.

White, J. 1979. The plant as a meta-population. Annual Review of Ecology and Systematics 10:109-145.

4

Canopy Development and Management of Guava for High Quality Fruit Production

K. K. Srivastava

Division of Crop Production, ICAR-Central Institute for Subtropical Horticulture, Rehmankhera, Kakori, Lucknow-226 101, Uttar Pradesh

The low productivity in guava because of traditional planting distance (6-7 m apart) with no canopy management practices. Guava under traditional and high density planting system, though it gives high yield during initial stage but as the tree advances in age light interception and aeration decreases inside the canopy due to shading and overcrowding, resulting low yields and poor quality. In order to develop efficient canopy architecture for HDP and traditional planting system, the ICAR-CISH has standardized training system and canopy management with objective to have sustained high yield and quality of guava.

Guava bears flower on current shoots produced on mature wood laterally or terminally. In general flowers are produced in the leaf axils in 1 or 2-3 in cluster. The crop load depend on number of new shoots, hence more new shoots resulting more fruiting surface area.

High Density Planting System

Accommodation of more plants per unit area is known as high density planting system. Since guava respond well to pruning and fruiting starts 40-50 days after pruning and more new shoots emerged on which more flowers and fruits appears, thus yield increased. For successful HDP suitable canopy shape and its management is prerequisite. Guava is usually planted 6-7 m distance accommodating 277 to 204 plants per hectare the yield vary 35-80 kg/tree with yield per hectare 15-12 t/ha. In traditional planting density usually fruiting takes place on high on the tree

Central Leader Training System

Guava is usually trained on free central leader training system in which main leader grow long (6-7 meters) with side branches at regular interval. Starting form 1-2 meter height from ground level. Most of the guava old guava varieties planted at 6-7 m distance on the farmer's field is on this training system. In this system the fruit zone lying on the upper periphery and branch breaking are the natural phenomenon. The yield in free standing central leader system varies 30 kg/tree to 150 kg/tree. Fruit color and quality are not appealing fruits area small and spraying etc. difficult job also need more spraying materials will make tree canopy large.

Photo: Central leader training system in high density planting system

Open Central Leader System

Due to precocious growing habit, the new guava orchards are now been planted on high density planting system. The grafted plants after planting is the field are headed back at 50-60 cm from ground and 4-5 primary branches are promoted at 10-15 cm interval with 45-60° crotch angle the upright growing branches are removed as such branch pruning. The primary scaffolds are headed back at 50-60 cm length from emergence of secondary 3-4 branches on which 2.0 tertiary branches are promoted for taking fruits. The central leader removed to make the tree shape as open bowl. Such system will allow good light interception in the inner most canopy layer too. Removal of water shoots are required to channelize the photo syntheses in required direction. The height of the open central leader trees are kept control by 60-70 pruning annually and thinning out is the regular phenomenon.

Photo: Open central leader training system

Photo: Pruning in Open central leader system

Espalier Training System

Espalier is a horticultural technique for controlling woody tree growth for production of fruit. In temperate climate espalier trained like wall to reflect more sunlight. The word espalier is French, which means something to rest on shoulder earlier referred only actual trellis on which a tree was trained to grow. The iron angles post having 9 feet length were fixed with the help of cement and concrete at 6-7 meters intervals on which 5-6 wires (tiers) were erected. First wire erected at 45 cm from ground level and remaining wires at 45-50 cm intervals. After erecting the structure well feathered grafted saplings were planted at 1.5x3.0 m (R-R x P-P) distance during July- August. In this system 2,222 trees per hectare are accommodated as compared to traditional system (204 plants/ha). First scaffold branches were trained at 45cm from ground level and rest scaffolds at 45-50 cm interval. Thus 8-10 main branches allowed on 4-5 tiers, on which fruiting shoots to be promoted. In order to promote side branch notching and girdling were done as required. In this training system top most scaffolds are at 260-270 cm or 8-9 feet height. The main trunk was terminated leaving 10-15 cm from last scaffolds, thus final tree height was 8-9 feet in espalier architecture, thickness of the. Total spread of tree canopy was 1.5-2 feet in both the sides (between rows).

Guava trained on espalier system are compact and its periphery is close to the main trunk, hence, effective cropping area is more, leading to a larger production. Since most of the leaves are exposed to sunlight, hence, maximum light interception took place. High dry matter production led to more photosynthates and attractive in appearance with red blushed fruit surface. Guava in this architecture bears 35 to 40 fruits (7-8 kg/tree) and projected yield (14.89 t/ha) from one year old tree or 35-45 kg per tree after 4 year of age.

Advantages of Espalier System

1. Ease in training, pruning, fruit picking and inter-culture operations.
2. Ease in spray of growth regulators, insecticide and fungicide.

3. Minimum wastage of spray materials and more trees can be sprayed in a shorter period of time.
4. Branch, shoot as well as leaf positioning are possible in this architecture.
5. Since the scaffolds are managed in tier system hence, good air drainage helps to minimise microbial population.
6. Cost effective and easy to develop this structure.

Photo: Espalier training system

Photo: Pruning in Espalier training system

Pruning in Guava

The bearing behavior of guava warranted for good response to pruning and training system. The fruiting takes place on 30-45 days old new shoots which emerged after pruning. The pruning is usually carried out in North India for Mrig bahar crop from 15th May to 15th June. Early crops can be harvested from early pruning. Pruning involved removal of one year old shoot or annual extension growth at 60-70% with 10-15% thinning out or complete removal of inner shoots or upright growing shoots. Annual pruning is required for growth of the new shoots for formation of flowering surface.

Crop Regulation for Encouraging Winter Season Guava

Guava bears fruits in two seasons (rain and winter) and most of the farmers prefer to take winter season fruits due to good taste and quality. Better winter season crop is achieved by creating stress to plants through withholding of irrigation and fertilization during March to June months. Pruning of 60-70 per cent of annual growth and thinning of shoots arising inside the canopy along with the removal of criss-cross branches, dead and dried wood should be done during the third week of May. The cut portion of twigs and branches should be pasted with copper oxy-chloride or Bordeaux paste. For avoiding the rainy season crop, stop fertilizer application as well as irrigation till the onset of monsoon. Digging of tree basin up to 6-8 inches depth will induce stress in the trees, which will result in good flowering by July- August for winter season crop.

Yield of quality fruits in guava under traditional planting distance and high density planting system can be sustained only regular pruning and canopy management system. In traditional planting system guava is trained on free standing central leader training system pruning and canopy management is pre-requisite.

5

Agro-Advisory Services for Management of Mango Insect Pests

Gundappa Baradevanal[1] and Singh H.S[2]

[1]*ICAR–National Bureau of Agricultural Insect Resources, Bengaluru, Karnataka*
[2]*ICAR- Central Institute for Subtropical Horticulture, Lucknow, Uttar Pradesh*

Introduction

Mangoes (*Mangifera* Indica L.), often referred to as the "king of fruits," are celebrated for their rich flavour, nutritional value, and versatility in culinary applications (Swaroop *et al*., 2018). Globally, mangoes hold a significant economic and cultural position. Mangoes are grown in over 90 countries, with India, China, and Thailand being leading producers. They contribute significantly to the agricultural economies of these and other tropical and subtropical regions (Makarabbi, 2023). Mangoes are a major export commodity for many countries, generating substantial foreign exchange. For instance, India's mango exports account for a sizable share of the global mango market, valued at millions of dollars annually. In many developing countries, mango farming supports millions of smallholder farmers, providing employment and a source of income (Thakor, 2019). Mangoes are enjoyed in numerous forms—fresh, dried, as juices, pickles, and chutneys. They are also a key ingredient in various traditional dishes and desserts across different cultures. Mangoes play a central role in several cultural festivals and rituals, symbolizing prosperity and good fortune in countries like India and the Philippines. Mangoes are rich in essential vitamins, minerals, and antioxidants, including vitamins A and C, folate, and fiber. Their consumption is associated with various health benefits, such as improved immunity, digestive health, and skin care (Yahia *et al*., 2023). While mangoes offer significant benefits, their cultivation faces several challenges, primarily due to pest infestations that can severely impact both yield and fruit quality. Effective management of these pests is crucial to ensure sustainable mango production (Singh and Baradevanal, 2021).

Pests and diseases can lead to significant economic losses, not just through direct damage but also by increasing costs associated with pest control and reduced marketability of affected fruits. Pest infestations lower fruit quality, making them unsuitable for export and processing markets, and can drastically reduce overall yield. Sustainability (Reddy *et al.*, 2018). Over-reliance on chemical pesticides can lead to environmental issues such as soil degradation, water contamination, and pest resistance. There is a growing need for integrated and sustainable pest management approaches (Baradevanal and Kumarnag, 2021). Effective management of mango pests requires a comprehensive understanding of pest biology, behavior, and the application of integrated pest management (IPM) strategies (Singh and Baradevanal, 2021). This chapter provides detailed guidance on identifying, monitoring, and controlling key mango pests, aiming to equip farmers with the knowledge and tools needed for successful mango cultivation and to maximize their economic and social benefits.

Importance of Agro-Advisory Services

Agro-advisory services play a critical role in modern agriculture by providing farmers with timely, relevant, and actionable information. These services enhance agricultural productivity and sustainability through various means. Farmers receive advice on optimal crop varieties, planting schedules, and cultivation techniques based on local agro-climatic conditions. This enables them to make informed decisions that maximize yields and quality (Raj and Garlapati, 2020). Advisory services offer timely information on pest outbreaks, disease symptoms, and effective management practices, helping farmers to mitigate crop losses and reduce reliance on chemical inputs. Guidance on efficient use of water, fertilizers, and pesticides contributes to sustainable resource management, reducing costs and environmental impact (Kumar, 2020). Advisory services provide accurate weather forecasts and climate information, enabling farmers to plan agricultural activities such as sowing, irrigation, and harvesting more effectively. Farmers receive advice on climate-resilient crop varieties and practices, helping them to adapt to changing climate conditions and reduce vulnerability to weather extremes (Amith *et al.* 2022). Agro-advisory services empower farmers and rural communities by enhancing their knowledge, skills, and overall capacity to engage in profitable and sustainable farming (Kilelu *et al.* 2022). The rise of mobile and internet-based agro-advisory services has transformed information delivery, making it more accessible and timelier. Farmers can receive real-time updates and interact with experts through digital platforms, enhancing their ability to respond to agricultural challenges promptly. Leveraging big data and analytics, agro-advisory services provide farmers with insights based on historical data, current trends, and predictive models, aiding in precise decision-making and planning (Sarkar *et al.* 2023). Agro-advisory services are vital for modern agriculture, offering essential support that enhances productivity, economic efficiency, and the overall empowerment of farmers.

Components of Agro-Advisory Services for Insect Pest Management

Agro-advisory services for insect pest management are comprehensive systems designed to provide farmers with the knowledge, tools, and support needed to effectively manage pest populations. These services integrate various components that work together to ensure sustainable and effective pest control. Here are the key components:

Pest Monitoring and Early Warning Systems : Advisories promote regular field scouting to monitor pest populations and identify early signs of infestation. Training on appropriate sampling methods for different pests, such as sweep nets, sticky traps, and pheromone traps, to assess pest densities and dynamics (Prasad and Rao 2022). Use of remote sensing technologies to detect pest hotspots and assess crop health over large areas (Luo *et al.* 2023). Mapping pest distribution and movement patterns to predict outbreaks and guide targeted interventions (Baradevanal *et al.* 2023). Development and promotion of mobile applications for farmers to report pest sightings and receive real-time alerts (Oteyo *et al.* 2021) Encouraging farmers to maintain logs of pest occurrences, control measures used, and their outcomes for better tracking and analysis.

Pest Identification and Diagnosis : Distribution of pest identification charts, handbooks, and mobile apps with images and descriptions of common pests and their symptoms. Provision of diagnostic kits including magnifying glasses, microscopes, and basic lab tools to help farmers identify pests accurately (Sehgal, 2023). Regular visits by extension officers to assist with pest identification and provide tailored recommendations (Wuepper *et al.* 2021). Establishment of local diagnostic centers where farmers can bring samples for expert identification and advice (Sharma, 2024).

Integrated Pest Management (IPM) Training and Support: Conducting educational programs on the principles and practices of IPM, including pest life cycles, natural enemies, and control strategies. Setting up model farms or plots to demonstrate effective IPM practices and allow farmers to observe and learn (Rejesus and Jones, 2020).

Information Dissemination Channels : Dissemination of pest management information and alerts through mobile apps and SMS services to reach farmers quickly and efficiently. Development of websites and online portals where farmers can access comprehensive resources on pest management, including best practices, research updates, and case studies (Rustia *et al* 2022). Organizing community-based training and knowledge-sharing sessions to foster peer learning and collective problem-solving. Use of local radio and TV channels to broadcast pest management tips, updates, and success stories to a wider audience. Distribution of printed materials with detailed information on pest identification, management strategies, and safety guidelines. Regular newsletters featuring updates on pest trends, control measures, and farmer experiences (Tonle *et al*. 2024).

Access to Resources and Inputs : Establishing connections with suppliers to ensure availability and access to effective pesticides, biocontrol agents, and other pest management inputs. Facilitating access to government subsidies, discounts, or bulk purchase options to reduce the cost of pest management inputs for farmers. Providing information on the proper use and maintenance of spraying equipment and other pest control tools. Advising on the use of personal protective equipment (PPE) to ensure farmer safety during pesticide application. Setting up helplines and help desks where farmers can seek immediate assistance and technical support for pest-related issues. Offering on-demand services such as pest identification, treatment recommendations, and pest management consultations.

Monitoring and Evaluation : Conducting regular surveys and collecting feedback from farmers to evaluate the effectiveness of advisory services and identify areas for improvement. Tracking key performance indicators such as pest incidence, yield improvements, and farmer satisfaction to measure the impact of pest management interventions. Collaborating with research institutions to develop new pest management strategies and technologies based on emerging trends and farmer needs. Implementing pilot projects to test innovative approaches and scale up successful models.

Agro-Advisory Services for Effective Insect Pest of Management in Mango

Almost a dozen of insect pests has been found damaging the crop to a considerable extent causing severe losses and, therefore, may be termed as major pests of mango. These are hopper, mealy bug, inflorescence midge, fruit fly, scale insect, shoot borer, leaf webber and stone weevil. Of these, insects infesting the crop during flowering and fruiting periods cause more severe damage.

Hoppers *Idioscopus clypealis, I. nitidulus* and *Amritodusatkinsoni*

Of all the mango pests, hopper is considered as the most serious and widespread. Large number of nymphs and adult insects' puncture and suck the sap of tender parts thereby reducing the vigour of the plants. They also secret a sweet sticky substance which encourages the development of sooty mould. A low population of hoppers is found throughout the year but it shoots up during February-April and June-August. Shade and high humidity conditions are favourable for their multiplication. Such conditions usually prevail in old, neglected and closely planted orchards.

Management

- Pruning, avoid high density planting and regulate flushes by reducing the inputs.
- *Verticillium lecanii* and *Beauveria bassiana* Coccinellid beetles, spiders, reduviid keep hopper population in check.
- Spraying of Lambda cyhalothrin 5% EC @ 1 ml/litre in mango orchard during off-season (November) especially on trunks.
- Integrating chemical insecticides with neem products depending on hopper density.

First spray: should be given at early panicle emergence with the following., At high hopper density (>4 hoppers/panicle) spray Imidacloprid 17.8% S L@ 0.3 ml/litre of water. At low hopper density (<4 hopper/panicle) spray Azadirachtin 3000 ppm 2 ml/ litre or 10,000 ppm 1 ml/litre .

Second spray: may be done with thiamethoxam 25% WG @ 0.3 g/litre at full length stage of panicles but before full bloom to avoid killing of pollinators. The third spray if required should be done at pea size fruit with Imidacloprid (as above)

Mealy bug *Drosicha mangiferae*

It is a major pest that causes severe damage to mango. Nymphs and adults suck the sap and reduce the vigour of the plant. They secrete honey dew which encourages the development of sooty mould. The female, after copulation, crawl down the tree in the month of April-May and enter in the cracks in the soil for laying eggs in large numbers encased in white egg sacs. The eggs lie in diapause state in the soil till the month of November - December. Just after hatching, the minute newly hatched nymphs crawl up the tree. After climbing up the tree they start sucking the sap of tender plant parts.

Management

- Modified trunk banding technique (Singh *et al.* 2022) should be used in November- December have been found effective barrier to stop the ascent of nymphs to the trees.
- Spraying with malathion 30% EC @ 1 ml/litre or profenophos 20% EC @ 2 ml/litrewith spreader or sticker (1 ml per litre) for killing the pest on foliage.
- The entomogenous fungus *Beauveria bassiana* @ 5 ml/litre of water was found to be an effective bio-agent in controlling the nymphs of the mealy bug.
- Flooding of orchards with water in the month of October kills the eggs and ploughing the orchards in the month of November exposes the eggs to sun's heat.
- Raking the soil around the tree trunk and mixing of 1.5 per cent Chorpyrifos dust @ 250g/tree was also found very effective in reducing the mealybug incidence.

Inflorescence midge *Procontarinia mangiferae (=Erosomyia indica)*

The midge damages the crop in three different stages. The first attack is at the floral bud burst stage then fruit set and tender new leaves encircling the inflorescence. The most damaging one is the first attack in which the entire inflorescence is destroyed even before flowering and fruiting. The flies lay eggs on inflorescence. Upon hatching, the minute maggots penetrate the tender parts and feed on them. The floral parts finally dry up and are shed. The mature larvae drop down into the soil for pupation. There are 3-4 overlapping generations of the pest spread over the period from January-March. Thereafter, as the weather conditions turn unfavourable, the mature larvae undergo diapause in the soil instead of pupating. They break diapause in following January.

Management

- As the larvae pupate in the soil, ploughing of the orchards expose pupating as well as diapausing larvae to sun's heat which kills them.
- Soil application of chlorpyrifos 1.5% dust kills pupating as well as diapausing larvae in the soil. The insecticide in the soil should be applied after monitoring larval population on white sheet below the tree.
- Spraying of Dimethoate 30% EC @ 1.5 ml/ litre of water at the bud burst stage of the inflorescence has been found effective in controlling the pest population

Fruit fly *Bactrocera dorsalis. B. zonata*

The oriental fruit fly is one of the most serious pests of mango in the country, which has created problem in the export of fresh fruits. *Bactrocera dorsalis* and *B. zonata*are the most common fruit flies. The female punctures the outer wall of the mature fruits with the help of its pointed ovipositor and insert eggs in small clusters inside the mesocarp of maturing fruits. After hatching, the larva feeds on the pulp of fruit which appears normal from outside, but drops down finally. The mature maggots fall down into the soil for pupation. The emergence of fruit fly starts from April onwards and the maximum population is recorded during May-July which coincides with fruit maturity.

Management

- Soil raking around and below trees to a depth of 6 cm. Twice – two weeks after start the fruit maturity and three weeks later. Ploughing may be done in winter.
- Collection and destruction of fallen fruits weekly starting from initiation of fruit maturity
- Male Annihilation Technique @ 10 blocks /ha starting from 45 days prior to fruit maturity. Replace the blocks after 30 days or even early.
- Bait Annihilation Technique (0.1% insecticide,10% jaggery or banana in water) starting 45 days prior to fruit maturity. Spray in spots of 40 ml at a rate of 200 spots (approximately 8 l/ha). Apply bait spray in spots to surroundings hedges also.
- Avoid delay in harvesting and, if needed give post-harvest hot water treatment within 24 hr. after harvest
- Hot water treatment or vapour heat treatment (VHT) of fruits before storage and ripening for killing the larvae can be done. After proper harvesting select uniform sized undamaged fruits. Dip them in 5%solution of sodium chloride in cold water for one hour. This will kill 95% eggs in fruit epicarp and also remove the externally present pesticide residues. Post-harvest immersion of mango fruits in hot water at 48 $\pm$ 1° for 45-60 minutes give 100% mortality of fruit fly eggs in the epicarp without affecting fruit quality

Scale insects: Scale insects have assumed the status of serious pest in certain parts of the country. *Pulvinaria polygonata, Aspidiatus destructor, Ceroplastis* sp. and *Rastococus* sp. are some of the most common scale insects infesting mango crop. The nymphs and adult scales suck the sap of the leaves and other tender parts and reduce the vigour of the plants. They also secrete honeydew which encourages the development of sooty mould.

Management

- Pruning of infested branches and burning
- Around 40 species of predators and parasites have been reported to attack the scale
- Sprays are only effective on the crawler stage of scales.
- Control is difficult on other life stages.
- Adults are firmly attached to the plant and remain so after their death that may give a false impression of the pest status. The application of pesticides may kill natural enemies of the scale and result in a resurgence of the pest

Bark-eating caterpillar *Inderbela* spp.

The old, shady and neglected orchards are more prone to attack by this pest. Larvae of this moth feed on the bark. The caterpillar spins brown silken web on the tree which consists of their excreta and wood particles. Larvae also make shelter tunnels inside the stem in which they rest. Larvae actually feed from April to December. There is onlyonegenerationinayear.

Management

- The caterpillars can be killed by inserting an iron spike into the tunnels.
- Mix chlorpyrphos 2 ml per litre of water and apply the bark eating caterpillar infested area with a brush at 15 days interval.

Stem borer *Batocera rufomaculata*

Grub of this beetle feeds inside the stems boring upward resulting in drying of branches. Eggs are laid either in the slits of tree trunk or in the cavities in main branches and stems covered with a viscous fluid. Pupation takes place within the stem. Beetle emerges in July-August. There is only one generation of the pest in a year.

Management

- Exclude alternative host trees from mango orchards and remove the dead trees and infested branches from the garden to prevent the spread of the pest.
- Mix chlorpyrphos 2 ml per litre of water and inject the solution in infested hole cover with mud. If infestation is severe then apply the copper oxychloride paste on the trunk of the tree to prevent disease incidence

Shoot gall psylla*Apsyllacistellata*

It is a very serious pest of mango in many parts of India, particularly in Terai region of U. P., north Bihar and West Bengal. This pest creates green conical galls in leaf axis. The activity of the pest starts from August. The galls dry out after emergence of psyllid adults in March. The females lay eggs in the midribs as well as in lateral axis of new leaves. Nymphs emerge from eggs during August-September and crawl to the adjacent buds to suck cell sap. As a result of feeding, the buds develop into hard conical green galls. The galls are usually seen during September-October. Consequently, there is no fruit set. There is only one generation of the pest in a year.

Management

- The galls with nymphs inside should be collected and destroyed to prevent carryover of the pest.
- The pest can effectively be controlled by spraying Dimethoate 30% EC @ 2 ml/litre or Quinalphos 25% EC @ 1.5ml/litre at 2-week intervals starting from the middle of August.

Thrips *Scirtothrips dorsalis*

Thrips have created havoc in most of the mango-growing areas of the country. Approximately 50 percent of damage of fruits by thrips was recorded in severely affected orchards. Thrips infestation starts with the new flushes and panicle emergence during the 13th to 22nd standard meteorological weeks (Second week of March to last week of May) in the northern plains. The larval and adult stages are the damaging stages. The damage on the mango young leaves, growing buds, inflorescence, flowers, immature and developing fruits by lacerating and sucking the sap from the tissues. This causes silvery or brown patches on the affected parts where the plant cells have been destroyed. As a result of its damage curling up of leaves and wilting of inflorescence were also recorded. In severe cases, affected fruits become rusty in appearance. This pest can damage the entire new growth, if not treated properly.

Management

- Monitor for thrips infestation by placing blue or yellow sticky traps at regular intervals.
- Neem-based pesticides control young nymphs effectively, inhibit the growth of older nymphs and reduce the egg-laying ability of adults. Spaying of Neem Seed Kernel Extract (5%) or Neem oil (2%) reduces the initial stages of the thrips effectively.

- If the infestation is severe, spray with insecticides like thiamethoxam 25% WG (0.3 g/lit) or Imidacloprid 17. 8% SL (0.3 ml/lit) Spinosad 45% SC (0.4 g/lit).

Frit borer: The fruit borer affects both mesocarps as well as seeds. The young larvae scrap the epidermal tissue and the mature ones bore into the fruit and feed on the pulp. The affected fruits exude resin and black or brown rough sticky patches can be seen on the infected area.

Management

- The spray of Lambda-Cyhalothrin 5% EC (1 ml per litre of water) or quinalphos 25% EC (1.5 ml per litre of water) at the marble stage of fruit and if needed repeat the spray after 15 days.

Leaf webber, *Orthaga euadrusalis* Walker

This pest has attained a serious and emerging problem as one of the most obnoxious pests in northern states. Old orchards with lesser space between the tree canopies harbour more insects than open orchards. Its infestation starts from April and goes up to December. Initially, caterpillars feed on the leaf surface by scrapping. Later they make a web of tender shoots and leaves together and feed within. The infestation is severe in shady conditions.

IPM Strategy

- Mechanical removal of leaf webs infested by leaf Webber by leaf web removing device and burning them.
- Spraying with lambda-cyhalothrin 5% EC @ 1 ml/litre or quinalphos 20% EC @1.5 ml/litre at 15 days interval

Shoot borer, *Chlumetia transversa* Walker

Larvae enter the mid rib of leaves and then enter into the young shoots through the growing points by tunneling downwards excrete out excreta through the entrance hole and the shoot becomes hollow. Larvae bore into the young shoots resulting in drooping of leaves and wilting of shoots. Larvae also bore into the inflorescence stalk

Management

Need based spray of quinalphos 20 % EC @ 1 ml /litre at fortnightly intervals during emergence of new flesh

Spray schedule for major insect-pests mango

Crop phenology	Pest targeted	Pesticides and its dose
Nov -Dec Jan- May	Mealy bug Mealy bug	Tree banding (technology may be obtained from CISH) If ascended on tree Dimethoate 30 % EC (1.5 ml/L)
Small panicle less than 1-2 inch, before flower opening (Feb –March)	Hopper	Imidachloprid 17.8% SL (0.3 ml/ L) or Thiamethoxam 25% WG (0.3 g/L) or Buprofezin 25 % SC (1 ml/ L) or Deltamethrin 2.80 % EC (0. 5 ml/L)
	Panicle/ flower midge	Dimethoate 30 % EC (1.5 ml/ L)
	Thrips	Imidachloprid 17. 8% SL (0.3 ml/L) or Thiamethoxam 25% WG (0.3 g/L) or Lamda cyhalothrin 5% EC (1 ml/L) or
After fruit set (March-May)	Hopper	Buprofezin 25% SC (1 ml/L) or Dimethoate 30 % EC (1.5 ml/L)
	Thrips	Imidachloprid 17.8% SL (0.3 ml/litre) or Thiamethoxam 25% WG (0.3 g/L)
Fruiting stage (May-June)	Semiloopers/ fruit borers	Dimethoate 30% EC (1.5 ml/L) or Lamda cyhalothrin 5% EC (1 ml/L)
	Fruit fly	Methyl eugenol-based traps (10 traps/ha), installation 30 days before fruit maturity
Maturity stage (June-July)	Fruit fly	Replacement of lure in Methyl eugenol-based traps (10 traps/ha).

Conclusion

Effective management of mango pests is crucial for enhancing yield quality and improving the livelihood of farmers. Agro-advisory services play a pivotal role in equipping farmers with timely, accurate, and location-specific information to manage pests sustainably. These services leverage a combination of traditional knowledge, scientific research, and modern technology, such as remote sensing, weather forecasting, and digital platforms, to address the dynamic challenges faced by mango cultivators. The integration of pest management practices with agro-advisory services ensures that farmers can implement preventive and curative measures at the right time. Techniques like Integrated Pest Management (IPM), combined with real-time alerts on pest outbreaks, can significantly reduce reliance on chemical pesticides, safeguarding both the environment and consumer health. Furthermore, community-based advisory services empower smallholder farmers by providing collective solutions and fostering knowledge-sharing. The success of agro-advisory services lies in their accessibility and adaptability. Ensuring that farmers receive information in their local language, through user-friendly channels like mobile applications, SMS alerts, or community radio, enhances their adoption. Partnerships between government agencies, research institutions, private organizations, and farmers' cooperatives are key to scaling up these initiatives and addressing region-specific pest challenges effectively.

References

Amith, G., Avinash, G., Haroli, M., Thimmareddy, H., & Dharani, C. (2022). Agromet advisory services for climate smart agriculture. Journal of Experimental Agriculture International, 44(4), 1-7.

Baradevanal, G., & Kumarnag, K. M. (2021). Biointensive Integrated Pest Management in Mango. In Biointensive Integrated Pest Management for Horticultural Crops (pp. 101-121). CRC Press.

Baradevanal, G., Chander, S., Singh, H. S., Reddy, D. S., & Rajan, S. (2023). Mapping the risk of quarantine pest Sternochetus mangiferae under different climate change scenarios through species distribution modelling. International Journal of Tropical Insect Science, 43(3), 919-932.

Kilelu, C. W., van der Lee, J., Koge, J., & Klerkx, L. (2022). Emerging advisory service agri-enterprises: a dual perspective on technical and business performance. The Journal of Agricultural Education and Extension, 28(1), 45-65.

Kumar, S. (2020). Agromet advisory services: tool to mitigate the effect of extreme weather events. In Sustainable Agriculture (pp. 149-172). Apple Academic Press.

Luo, Y., Huang, H., & Roques, A. (2023). Early monitoring of forest wood-boring pests with remote sensing. Annual Review of Entomology, 68(1), 277-298.

Makarabbi, G. (2023). An analysis on performance of mango production in India. Asian Journal of Agricultural Extension, Economics & Sociology, 41(10), 968-976.

Oteyo, I. N., Marra, M., Kimani, S., Meuter, W. D., & Boix, E. G. (2021). A Survey on Mobile Applications for Smart Agriculture: Making Use of Mobile Software in Modern Farming. SN computer science, 2(4), 293.

Prasad, T. V., & Rao, M. S. (2022). Pest Survey and Surveillance for Pest Forecasting and Pest Management. Adaptation Strategies for Pest Management in Climate Change Scenarios, 43.

Raj, S., & Garlapati, S. (2020). Extension and advisory services for climate-smart agriculture. Global climate change: Resilient and smart agriculture, 273-299.

Reddy, P. V. R., Gundappa, B., & Chakravarthy, A. K. (2018). Pests of mango. Pests and their management, 415-440.

Rejesus, R. M., & Jones, M. S. (2020). Perspective: enhancing economic evaluations and impacts of integrated pest management farmer field schools (IPM-FFS) in low-income countries. Pest Management Science, 76(11), 3527-3536.

Rustia, D. J. A., Chiu, L. Y., Lu, C. Y., Wu, Y. F., Chen, S. K., Chung, J. Y., Hsu, J. C. and Lin, T. T., (2022). Towards intelligent and integrated pest management through an AIoT-based monitoring system. Pest Management Science, 78(10), 4288-4302.

Sarkar, N. C., Mondal, K., Das, A., Mukherjee, A., Mandal, S., Ghosh, S., . . . & Huda, S. (2023). Enhancing livelihoods in farming communities through super-resolution agromet advisories using advanced digital agriculture technologies. Journal of Agrometeorology, 25(1), 68-78.

Sehgal, M. (2023). Use of Information and Communication Technology (ICT) in Integrated Pest Management (IPM). e-SOUVENIR cum Compendium of, 1.

Sharma, M. Entomological Consultants: Clinical, Diagnostic, and Management Service. In Skill Development and Start-Ups in Entomology (pp. 11-33). CRC Press.

Singh, H. S., & Baradevanal, G. (2021) Insect pests and their integrated management strategies. Indian Horticulture, 66(4).

Singh, H. S., Baradevanal, G., & Beer, D. (2022). Modified tree trunk banding technology for mango mealybug, Drosichamangiferae (Green) management: A techno-economic analysis. Pest Management in Horticultural Ecosystems, 28(2), 133-137.

Swaroop, A., Stohs, S. J., Bagchi, M., Moriyama, H., & Bagchi, D. (2018). Mango (Mangifera indica Linn) and anti-inflammatory benefits: versatile roles in mitochondrial bio-energetics and exercise physiology. Functional Foods in Health and Disease, 8(5), 267-279.

Thakor, N. J. (2019) Indian mango–production and export scenario. Peach 18(107), 12.

Tonle, F. B., Niassy, S., Ndadji, M. M., Tchendji, M. T., Nzeukou, A., Mudereri, B. T., Senagi, K. and Tonnang, H. E. (2024). A road map for developing novel decision support system (DSS) for disseminating integrated pest management (IPM) technologies. Computers and Electronics in Agriculture, 217, 108526.

Wuepper, D., Roleff, N., & Finger, R. (2021). Does it matter who advises farmers? Pest management choices with public and private extension. Food Policy, 99, 101995.

Yahia, E. M., de Jesús Ornelas-Paz, J., Brecht, J. K., García-Solís, P., & Celis, M. E. M. (2023). The contribution of mango fruit (Mangifera indica L.) to human nutrition and health. Arabian Journal of Chemistry, 16(7), 104860.

6

Major Issues in Disease Management for Agro-Advisory in Subtropical Fruit Crops

P.K. Shukla

ICAR-Central Institute for Subtropical Horticulture, Rehmankhera P. O. Kakori, Lucknow-226 101, Uttar Pradesh

Among subtropical fruit crops mango, guava, bael and aonla are among the most important ones. The productivity of these crops is affected by many factors, among which diseases have their own economic significance. The crops use to suffer from a large number of diseases, however, economically important diseases having significance in agro-advisory are mentioned below.

Mango Wilt and Decline

The mango wilt disease caused by *Ceratocystis fimbriata* is one of the most devastating diseases (Shukla *et al.,* 2015, 2018b, 2020; Shukla, 2020). Association of *Berkeleomyces basicola and Lasidiplodia theobromae* in wilt, decline and branch drying has also been observed.

Symptoms: *Ceratocystis* infection on the trees causes dark staining of the affected wood, exudation of gum, leaf drying and ultimately death of the tree (Al Adawi *et al*., 2003; Shukla *et al.,* 2018b). Later symptoms include vascular discolouration beneath the gummosis (Masood *et al.,* 2008; Shukla *et al.,* 2018b). The symptoms may appear in the form of sudden wilt, decline or branch drying and accordingly the whole tree or branch may die within a period of a fortnight to six months of appearance of first symptom.

Wilt

Decline

Trunk infection

Root infection

Factors affecting disease development: The pathogen is closely associated with infestations of the wood-boring scolytid beetle *Hypocryphalus mangifera*, which is considered to spread the pathogen (Gunadappa *et al.*, 2016). Ambrosia beetles (*Xyleborus* and *Hypocryphalus* species) are attracted to diseased plants and produce large amounts of fine wood shavings when creating breeding galleries in the trunk and branches (Goitia and Rosales 2001). These wood shavings and faecal material are pushed outside the tree as the galleries are excavated, and the frass contains spores and fragments of mycelium that may be blown in the wind. Mango trees may be infected through the roots from soil borne inoculum or by direct infection in the stem (Shukla *et al.*, 2018b). Pruning wounds are common entry points for the fungus. Because there may be extensive mycelial growth within a plant before symptoms appear, propagative cuttings may be an effective method of dispersal. Uncontrolled irrigation, intercropping, wounding of trees, improper fertilizer application predisposes the mango trees to wilt (Shukla *et al.*, 2018b).

Management: Recently, Shukla (2020) developed an integrated management package of practices against mango wilt. It included:

1. Minimum tillage near tree trunk (1.0-2.5 m) and under canopy deep ploughingmust be avoided to minimize root damage.
2. The root zone of trees in wilt affected orchards should be drenched with 0.1% solution of thiophanate methyl 70WP or azoxystrobin 11 + tebuconazole 18.3 SC or hexaconazole 5 SC @ 25-50 l/m^2.
3. Cut ends should be pasted with copper sulphate or copper oxychloride 50 WP @ 5.0% after every pruning followed by spray with carbendazim 50 WP or propiconazole 25 EC or azoxystrobin 11 + tebuconazole 18.3 SC or hexaconazole 5 SC @ 0.1% after pruning or on the appearance of the first symptom.
4. Infestation of Scolytid beetle may be managed by the spray of chlorpyriphos 20 EC @ 0.2 per cent at 7-15 days interval.

5. Irrigation must be done using a drip system or channels to avoid disease dispersal through water from an infected tree to healthy trees.
6. Recommended doses of manure and fertilizers should be applied in orchards and proper soil moisture should also be maintained regularly to promote recovery of tree.

Mango Powdery Mildew

Powdery mildew, caused by the fungus, *Oidium mangiferae,* is a weather dependent devastating disease of mango that cancause loss up to 90 percent if enough inoculums are present during susceptible stage of the crop (Sinha *et al.*, 2001; Misra, 2001; Shukla *et al.*, 2016b).

Symptoms: The symptom of powdery mildew appears as the white powdery superficial growth of fungus mainly on inflorescences, but it may also be seen on youngleaves and tiny fruits. The infected flowers and fruits may drop down and severe losses may occur.

Initial infection powdery mildew

Powdery mildew after flower drop

Factors affecting disease development: The infection takes place within 5-7 hours after conidial germination, and conidia are produced within 5 days of infection under favorable conditions (Johnson, 1994). The most suitable minimum and maximum temperature for disease development has been reported 10-13°C and 27-31°C respectively. High humidity favours infection and low humidity disease spread. Dry weather with maximum temperature between 29-34°C favours the faster development of the disease.

Disease Management

If incidence is seen on more than 10 percent panicles under suitable weather comditions, spray with tebuconazole + trifloxystrobin @ 0.05%, hexaconazole @ 0.1% or wettablesulfur @ 0.2% should be done.

Mango Blossom Blight

Blossom blight disease can drastically reduce fruit set if panicles and flowers are severely infected under favorable conditions.

Symptoms: The infection has been recorded since inception of the panicles, continues during their development and flowering, if weather favours. The symptoms appear in the form of blackish brown specks or spots on the peduncle. The black colour spots are developed on flowers, gradually enlarge, infected flowers becomes black and infected flowers fall off. The severity of the disease depends up on the weather.

Causal organism:*Alternaria alternata* and *Colletotriculum gloeosporioides* have been found the major cause the disease (Lonsdale and Kotze, 1993). Rawal and Venugopalan (2011) established positive correlation between severity of blossom blight and brightness of sunshine, rainfall, rainy days and vapour pressure deficit.

Management: The disease can be managed by the spray of carbendazim + mancozeb (12+63%) formulation @ 0.2 per cent or trifloxystrobin + tebuconazole (25+50%) @ 0.05 per cent or hexaconazole 5SC @ 0.1 per cent.

Mango Leaf Anthracnose

Leaf anthracnose of mango,caused by *Colletotrichum gloeosporioides,* is responsible for significant losses to new flushes in the form of leaf spot, twig blight, wither tip, especially after rainy season. High relative humidity (>80%), heavy dew and rain favour rapid disease development. Brown todark brown round spots develops on the leaves; the spots grow rapidly under favorable conditions, which later rupture to create typical shot hole in leaves. The infected petiolesbecome grey or black and dry. The graft union use to suffer from anthracnose infection and the infected plants usually dies.

Management: Fresh infection can be reduced by minimizing the sources of inoculums in the orchards. Diseased plant partson small trees or on ground should be removed and disposed properly. Disease can be managed by the spray of copper oxychloride 50WP @ 3g per litre of water or carbendazim 12 + mancozeb 63 WP @ 2 g per litre of water or trifloxystrobin + tebuconazole (25+50%) @ 0.5 g per litre of water as a precautionary measure or on appearance of first symptoms.

Mango Fruit Anthracnose

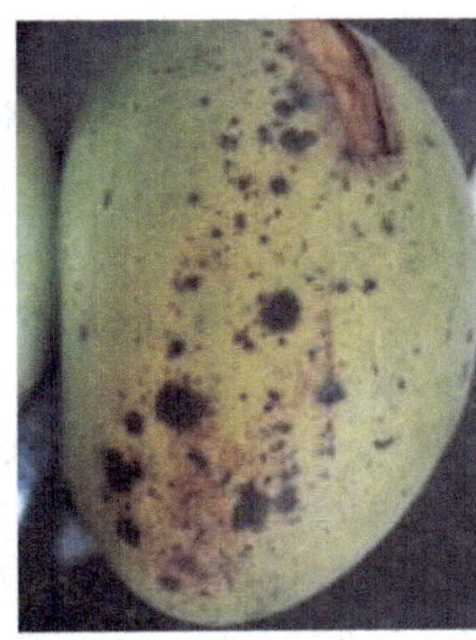

It is the most serious postharvest disease of mango fruits, particularly when rain and high humidity prevails during maturity periods (Shukla *et al.*, 2016a, 2017). Fruits at all growth stages are prone to the disease and may express typical disease symptoms (Yenjit *et al.*, 2004). Sunken necrotic lesions on the fruit surface and mummified fruits are typical symptoms (Rivera-Vargas *et al.*, 2006). The anthracnose of fruits becomes severe during ripening on tree or during transit and storage, which reduces the number of marketable fruits resulting in huge losses. Diseased fruits have deep brown to black round spots, which later coalesce to form large irregular blotches. The disease incidence is enhanced by rains and cloudy weather under field conditions; and high humidity and high temperature during storage.

Management

The integrated orchard management at critical phenological stages of crop is necessary round the year for postharvest management of anthrcnose (Ploetz, 2009). Inoculums build-up in orchard can be minimized by taking care of each new flush. The pre-harvest application of fertilizers, manures, growth regulators, canopy management influences the fruit physiology, chemical composition and extent of infection in fruits (Prakash, 2004). Plant resistance is the best and environmentally safe approach however, all commercial cultivars are susceptible to disease but susceptibility vary among them (Akem, 2006).

The inoculums build-up can be minimized by the spray of carbendazim, propiconazole, thiophanate methyl, tricyclazole and difenoconazole have also been found effective (Agrios, 2005; Shukla *et al.*, 2018a). Spray of mancozeb in mango orchard reduced anthracnose incidence on leaves (Islam *et al.*, 2016).

The hot water treatment (HWT) which includes dipping of fruits in hot water at 50-55°C up to 5 minutes (Alemu, 2014a,b) just after harvesting, is an age old and successful practice that can alone efficiently reduce anthracnose development on fruits (Uddin *et al.*, 2018). In case of high disease pressure, the potency of HWT can greatly enhanced by combining with fungicides (Kapse *et al.*, 2009; Kefialew and Ayalew, 2008). Different products of plant origin and essential oil *viz.*, castor oil, eucalyptus oil, lemongrass oil, garlic bulb extract, *Azadirachta indica*, *Zingiber officinale*, *Lantana* and *Curcuma longa* leaf extract have been found effective in suppressing the anthracnose and postharvest rots of mango fruits (Singh *et al.*, 2003; Duamkhanmanee 2008; Alemu *et al.*, 2014b).

Mango Stem End Rot

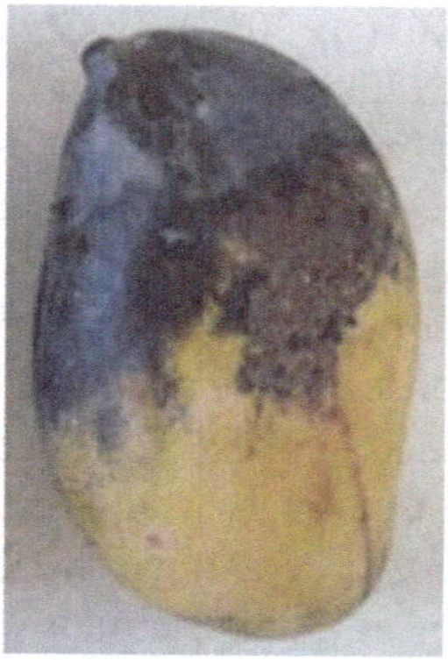

It is another economically serious postharvest disease of mango fruits worldwide (Yang *et al.*, 2012). The disease incidence is more severe in orchards having older trees and if fruits are subjected to prolonged storage (Cooke *et al.*, 2009). Several fungi *viz.*, *Lasiodiplodia theobromae, Aspergillus niger, Phomopsis mangiferae* or *Dithiorella dominicana* and *Colletotrichum gloeosporioides* have been isolated from diseased fruits (Cooke *et al.*, 2009). Among these, *L. theobromae* has been considered as the most important. The pathogen remains dormant on the unripe fruit. As the ripening process begins, fruit start rotting near stem endand becomes brown-grey, soft and rotting gradually covers the entire fruit surface (Govender, 2005; Prakash, 2004; Prakash *et al.*, 2011).

Management: Disease incidence can be managed by reducing inoculum load and infection in twigs by adopting round the year orchard management. Application of *Bacillus licheniformis* and covering fruits with polyethylene or paper bags was also found effective in reducing the incidence of stem end rot (CABI, 2005). Preharvest spray of thiophanate methyl, carbendazim, tebuconazole + trifloxystrobin, azoxystrobin were found effective in managing postharvest stem end rot (Khanzada *et al.*, 2005; Syed *et al.*, 2014). Singh (2011) applied hot water at 45°C or 50°C alone and in combination with sodium bicarbonate and propiconazole dip for 5 minutes each on artificially inoculated fruits and 83.79-96.41 per cent reduction in disease development was recorded over control. Preharvest spray of propineb and difenoconazole followed by hot wter treatment at 52 ºC for 10 minutes has also been found effective.

Mango Shoulder browning: The first symptom of the shoulder browning disease appears on the shoulder area of fruits in the form of deposit of fungal spores and mycelium of light grey, dark brown or black colour. The deposit on fruits become darker and covers most of the fruit surface as the rains proceed. No variety of mango is free from the disease; however, late maturing cultivars suffer severe losses from this disease. After 7-8 rainy days and more than 70 mm rainfall disease takes severe from (Shukla *et al*., 2016a).

Management: The disease can be managed by bagging of fruits (8×12 inch) before onset of monsoon or spray of propineb @ 0.2% or difenoconazole @ 0.05% up to 21 days prior to harvesting of fruit (Shukla *et al.*, 2018a). Round the year proper care of orchard and management of sucking pests also helps in reducing the disease severity.

Mango Dieback/ Twig Drying

Symptoms of dieback and twig drying/blight mostly appear during and after rains. Drying of twigs from tip to downward is the typical symptom of mango dieback disease. Sometimes, twigs may dry with gummosis from bark. Leaves of infected twigs become dry and internal tissue of the twigs become brownish. Such symptoms are mostly seen in orchards during post monsoon period.

Management: Pruning of infected twigs approximately 10-20 cm below dried portion, followed by cut ends pasting with copper oxychloride @ 5.0% and /or spray of copper oxychloride 50WP @ 0.3% (i.e.3.0 g per litre of water).

Internal necerosis of mango fruits: The disease is caused due to the deficiency of boron, which can be managed by the spray of solubor or folibore @ 2 g/ litre of water during fruit development stages.Soil application of borax may also be done @ 50-100 g per tree during rainy or post rainy season.

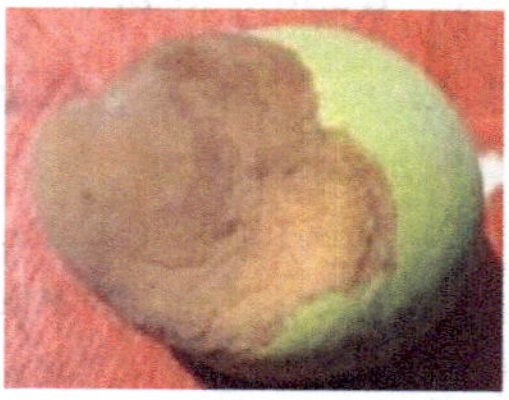

Guava wilt/ Decline

The guava decline is one of the most important diseases caused by the root-knot nematode, *Meloidogyne enterolobii*. Survey conducted for the past five years has shown infestation in over 50% of orchards (Khan *et al*., 2019). Nurseries have been the source for continuous spread of the nematodes to newer areas. The infestion of nemaodes enhances the infection of wilt fungi, *Fusarium* spp. The nematode infected plants may show symptoms of decline at very early age in the form of stunting, yellowing and shedding of leaves. It is well proven that if an orchard of guava is infested with nematodes, the economic return goes to the bottom first due to low productivity and next due to cost of nematode management.

Gauava wilt

Guava decline

Galling on roots

Management: The nematode infestation in guava orchards is challenging because it cannot be eradicated. The treatments available for its management e. g. fluopyrum, a chemical, is highly effective, but it is costly and its effect ends within six months; thus repetition becomes necessary. The bio-control agents like *Trichoderma harzianum, Pochonia chlamydosporia, Purpureocillium* *lilacinum, Bacillus* amyloliquefacies are effective but need frequent repetition. The uses of nematode resistant crops in inter cropping and organic products have been found moderately effective. Intercropping with nematode resistant crops has been found helpful (Khan and Misra, 2003)

Healthy planting material production is necessary to avoid introduction of nematodes along with the planting material. The soil and roots of nematode infected grafts must be treated with Fluopyrum 0.05% solution at 15 days prior to transplantation in main field. Field planting should be done as deep as possible with drenching of 0.05% solution of fluopyrum @ 2 litres per plant. The selection of field for establishment of guava orchard is also important as heavier soils are suppressive to nematodes. The best way is to assure absence of root-knot nematode in the field selected for establishment of a new guava orchard and also constant application of bio-agents.

Guava Anthracnose

The disease has great economic significance because it not only causes infection of fruits but also shoots. It is well known to cause pre- and postharvest losses to fruits. It is caused by *Colletotrichum gloeosporioides* (teleomorph: *Glomererella cingulata*) or *Gloeosporium psidii* which mostly causes the infection in each tender part of the plant (Sridhar and Sinha, 1996). Rapid disease development has been recorded at temperatures around 30 °C and supported by prolonged warm-wet weather. The extent of disease severity depends on the number of rainy days and amount of rainfall in a specified period. Injury to fruits enhances the chances of infection.

Symptoms: Since, the fungus causes infection of every tender portion of plant, it may cause dieback of growing tips and spots on leaves, flowers and fruits. The dieback symptoms develop if young shoots are infected (Sridhar and Sinha, 1996). Initial infection of the growing tip can be observed as tiny brown spots that grow fast under favourable conditions. The shape of spots remain round till they fuse to make larger patches and turn black. The growing shoot may be killed in the process. Similar symptoms also develop on leaves and under severe cases leaves may drop down (Sridhar and Sinha, 1996). Infection on fruit is very common during monsoon. Severely infected fruits do not grow further and become hard. If mature fruits are severely infected, they rot and drop down. Infection of flowers causes their blight and death.

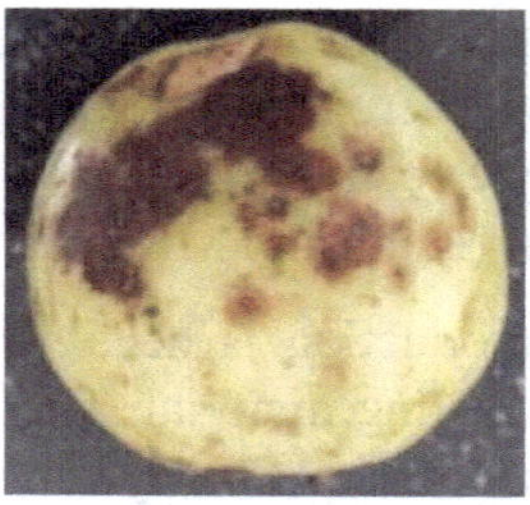

Fruit anthracnose symptoms

Management: Pruning for canopy management followed by removal of all infected leaves, fruits and branches reduces the disease. Butt *et al.* (1995) reported efficacy of copper oxychloride against the formation of acervuli and boric acid against mycelial growth of the fungus. Spray of thiophanate methyl @ 0.1 percent or carbendazim + mancozeb @ 0.2 percent or propineb @ 0.2 percent just after initiation of disease or preventive spray of fungicide followed by bagging of fruits has been found highly effective when done during early development stages of fruits.

Guava Stylar End Rot

Stylar end rot symptoms

The disease, caused by *Phomopsis psidii,* has been recorded for a long time (Sridhar and Sinha, 1996) but its increased severity has been recently recorded in normal as well as high density plantings, and thus getting higher economic significance. Initially light brown discoloration at the distal end of fruit is seen, which gradually progresses and becomes dark in colour. The infection also proceeds inside the fruit and rotting of pulp takes place (Sridhar and Sinha, 1996). Cracking of fruits due to boron deficiency followed by infection of fungus has also been recorded.

Management: Proper canopy management to harness maximum solar radiation, spray of solubore @ 2 g per litre of water followed by bagging of fruits for the rainy season crop has been found effective against the disease in farmers' fields. Repeated spray of zineb @ 2 g or guaiacol @ 0.025 g or α-naphthol @ 0.005 g per litre of water provides satisfactory reduction in disease severity. Postharvest rot can be minimized by dipping fruits in neem leaf extract and keeping fruits at <10 °C at low relative humidity (El-Eryan *et al.*, 2017).

Guava Fruit Canker

Fruit canker

The disease is reported to cause severe losses throughout the world in heavy rainfall areas (Lim and Manicom, 2003). The disease is caused by *Pestalotiopsis psidii* mostly during the rainy season (Sridhar and Sinha, 1996; Lim and Manicom, 2003) and frequently found associated with guava leaves and fruits in Uttar Pradesh. The maximum sporulation of the fungus occurs at 28-30 °C with RH above 96 percent. The fungus produces ascospores perithecium and conidia in black colour circular pycnidia are produced on fruits, which release conidia at maturity.

Symptoms: Immature fruits and leaves are commonly infected. The initial symptoms appear as tiny brownish or orange spots on fruits. The fully developed spots or cankers have raised periphery and depressed in center. Cracking of fruits may also be evident in severe cases, otherwise fruits remain underdeveloped. The infestation of bugs enhances the infection and severity of disease (Lim and Manicom, 2003). Symptoms on leaves may appear as orange-brown lesions.

Management: Disease can be managed by adopting measures similar to anthracnose and Phytophthora fruit rot. Pandey *et al.* (1983) advocated the postharvest dip of fruits in the extracts prepared from leaves of neem (*Azadirachata indica*) and Indian basil (*Ocimum sanctum*) for managing the rot.

Fruit Drop of Bael

Untimely fall of bael fruits happens throughout the fruiting period. It may be caused due to pathological and physiological reasons; or due to chilled injury during severe winter. Several fungi belonging to *Fusarium*species complex, causes infection in stem end portion as well as in fruits; are responsible for pathological fruit drop. The cracking of bael fruits has also been observed during post monsoon period. Serious efforts have been made to manage it, but the desired level of success, particularly against chilled injury, has not been achieved yet. The cultivar bearing over 1.0 kg weight fruits suffer more than

smaller fruit cultivars during winter. The fruits achieve physiological maturity desired for processing of fruits during October month, and if fruits are picked and processed till December month, the baelfarming remains profitable.

Management: The physiologicalfruit drop can be minimized by maintaining adequate soil moisture by irrigating the root zone of trees and spray with nepthlene acetic acid @ 100-200 ppm. Spraying of fungicides (Carbendazim + Mancozeb 2 g or Tebuconazole 1 ml or Propineb 1 ml or Crasosim-methyl + Chlorothalonil 1 ml per liter of water) and Potassium Sulphate 3-4 g per liter of water 2 to 3 times at an interval of 15 days has been found helpful in minimizing pathological fruit drop.

Aonla Rust

Rust of anola caused by *Ravenelia emblicae* is an economically important disease affecting fruits and leaves (Misra and Prakash, 1996). Its moderate to severe incidence occurs almost every year resulting in defoliation and damage to fruits.

Symptoms: The rust pustules develop on fruits are brown to black in colour, initially small in size but soon they grow. Under severe incidence of the disease, pustules coalesce and the major area of fruits is covered (Jat *et al.*, 2004). The infected fruits lose their aesthetic value. Pink to purple colour rust pustules develop on leaves. Under severe infection, heavy leaf fall has been recorded.

Symptoms of Rust on Aonla Leaves

Management: Since, differences have been observed in susceptibility of different cultivars of aonla and variation in incidence during years, therefore, decision towards management should be taken in view of possible damage. First spray should be done with wettable sulphur @ 0.4 percent with the appearance of first symptoms (Misra and Prakash, 1996). If necessary, the same fungicide may be repeated or hexaconazole @ 0.1 percent or chlorothalonil @ 0.2 percent may be used as an alternative. Three spray with bio-control agents *viz. Trichoderma viride* and *Pseudomonas fluorescens* were also found effective in managing the disease (Kumar *et al.*, 2017).

Aonla Blue Mould

Occurrence of blue mould (*Penicillium citrinum*) is very common wherever aonla is grown (Misra and Prakash, 1996). The disease is extremely important if fruits are injured during picking and handling.

Symptoms: The fungus causes water-soaked areas and brownish patches on the epicarp of fruit, which later changes the colour to turn yellow, brown and blue-green. Yellow exudates may also be seen on the surface of infected portions, which smell foul (Misra and Prakash, 1996). All affected fruits finally have the blue-green growth of the fungus on the entire surface.

Management: The rot can be minimized by careful handling of fruits to avoid injuries. Store rooms should be treated with ozone or nitrogen trichloride gas (Misra and Prakash, 1996). Fruits can also be treated with sodium chloride to avoid fungal infestation. Dipping of fruits in 0.1 percent solution of thiophanate methyl may also be recommended but such fruits may have residual toxicity. Fruits can be smeared with mint oil, which checks the rot.

Aonla Fruit Soft Rot

The disease caused by *Phomopsis phyllanthi* is not important if fruits are harvested with proper care. It has been observed to be more prevalent during winter (Misra and Prakash, 1996).

Symptoms: Initial symptoms come up as water soaked grayish-brown circular lesions on the fruit surface. Within 5-7 days the entire fruit surface may be covered and the colour of lesions becomes darker. Direct penetration of fungus has not been recorded and injury to fruits is the basic reason behind the development of disease.

Management: Utmost care should be taken in picking and handling of fruits. Harvested fruits should be treated in solution of thiophanate methyl @ 0.05 percent or mancozeb-carbendazim formulation @ 0.1 percent.

Aonla Fruit Rot

Postharvest fruit rot is a very common and severe problem in aonla (Misra and Prakash, 1996). The fruits are either infected while on tree or infection takes place after the harvesting. The aonla is a very delicate fruit, which is injured with little carelessness and injured fruits are infected or injured dead tissue are colonized by several fungi. As a result of which severe rotting of fruits takes place.

Phoma emblicae is responsible for dry rot of fruits (Misra and Prakash, 1996). Sometimes freshly harvested fruits were also found affected with the pathogen. The infected areas were observed light brown and measured from 1.0 to 2.0 cm in diameter but subsequently became depressed and turned black. Occasionally pycnidia were observed on the infected fruits.

Fruit rot caused by *Nigrospora sphaerica* is common during the month of December (Misra and Prakash, 1996). The fungus is common in nature but its loss potential to aonla is <10 per cent.

The losses due to dry fruit rot caused by *Cladosporium tenuissimum* and *C. cladosporioidies* has been estimated around 2-5 percent. Infection of *C. tenuissimum* has been observed during winters in north India (Misra and Prakash, 1996). Infection on fruits can be identified as lesions with light colour and loss of normal hardness in the infected portion. The fully developed circular lesions occupy a major area of fruits in the later stages. The symptoms of rot caused by *C. cladosporioidies* appear as amber colour irregular lesions. These symptoms are mostly seen sometime after harvesting of mature fruits and rarely on trees.

Pestalotia cruenta infection may also cause rotting of fruits mostly during the month of November (Misra and Prakash, 1996). The symptoms initially appear as brownish lesions on the fruit epicarp, which later cover a larger area of fruits. The pulp below the lesion may have discolouration and rotting. Infected areas may also have cottony growth of *P. cruenta* under favourable conditions.

Fruit rot caused by *Alternaria alternata* has mostly been recorded in dropped fruits (Misra and Prakash, 1996). The symptoms of rot initially appear as tiny brown colour spots and later enlarge in round shape. The brown colour of spots later becomes black and large lesions develop after joining neighboring spots. The center of such lesions softens at maturity of fruits.

Symptoms of Fruit Rot of Aonla on Tree

Management of fruit rots: Pre-harvest spray with thiophanate methyl or carbendazim (1 g/l) protects fruits from various types of infestations and infections. It must be done more than 2 weeks prior to harvesting of fruits so that residues remain at minimum. Fruit injury is the major issue in aonla, therefore, utmost care in harvesting is necessary. Harvested fruits must be kept in properly clean plastic baskets and be maintained at 10 °C before consumed (Misra and Prakash, 1996). The fruits may also be treated with sodium chloride to minimize infection and losses.

Citrus Canker

Citrus canker is an important and very common disease of citrus that appears in the form of raised light brown colour spots on leaves and fruits. The incidence of disease becomes severe under prolonged wetness of leaves and it spread at faster pace during rains.

Management: Incidence can be kept below significant loss limit by keeping canopy open through pruning to facilitate proper sunlight inside the canopy. Spray of a suitable antibiotic @ 0.2 g/ L of water or copper oxychloride 50 WP @ 3 g per litre of water may be done for its management.

References

Agrios GN (2005) Plant Pathology. Edn 5, Elsevier Academic Press, San diego, 329-347.

Akem CN (2006) Mango anthracnose disease: Present status and future research priorities. JPP 5(3): 266-273.

Al-Adawi, A. O., Deadman, M. L., Al Rawahi, A. K., Khan, A. J. and Al Maqbali, Y. M. 2003. Diplodia theobromae associated with sudden decline of mango in the Sultanate of Oman. Plant Pathology, 52: 419.

Alemu K, Ayalew A, Woldetsadic K (2014a) Effect of aqueous extracts of some medicinal plants in controlling anthracnose disease and improving postharvest quality of mango fruit. Pers Gulf Crop Prot3(3): 84-92.

Alemu K, Ayalew A, Woldetsadic K (2014b) Evaluation of antifungal activity of botanicals for postharvest management of mango anthracnose (*Colletotrichum gloeosporioides*). Int J Life Sci8(1):1-6.

Butt, A. A., Nasir, M. A. and Bajwa, M. N., 1995. In vitro evaluation of different chemicals against Gloeosporium psidii, the cause of anthracnose of guava. Pak. J Phytopath., 7 (1): 92-93.

CABI (2005). Crop Protection Compendium, 2005 ed. Wallingford, UK: CAB International.

Cooke T, Persley D, House S (2009) Diseases of fruit crops in Australia. CSIRO publishing, Australia, pp 157-173.

Duamkhanmanee R (2008) Natural Eos from lemon grass (Cymbopogon citrates) to control postharvest anthracnose of mango fruit. Int J Biotech 10(1): 104-108.

El-Eryan, E. E., Tarabih, M. E. and EL-Metwally, M. A., 2017. Influence of natural extracts dipping to maintenance fruit quality and shelf life of Egyptian guava. J. Plant Production, Mansoura Univ., 8(12): 1431-1438.

Goitia, W. and Rosales, C. J. 2001. Relacion entre la incidencia de escolitidos y la necrosis del cacao en Aragua, Venezuela. Manejo Integrado de Plagas, 62: 65-71.

Gundappa, Adak, T. and Shukla, P. K. (2016). Status of mango wilt vector, Hypocryphalus mangiferae in India. Indian Phytopath. 69 (4s): 361-363.

Islam MM, Hassan ME, Hossain SMM, Hasan MM, Islam NB (2016) Prevalence and management of some major diseases of mango at Dijapur district in Bangladesh. Int J ExptAgric 6(2):1-11.

Jat, R. G., S. K. Goyal and R. P. Didel (2004). Management of aonla rust incited by Ravenalia emblicae var. fruticolae Syd. Asian J. Bio. Sci., 8(1): 117-119.

Johnson, G. I. (1994). Powdery mildew. In: Compendium of Tropical Fruit Diseases (eds. R. C. Ploetz, G. A. Zentmyer, W. Nishijma, K. Rohrbach and H. D. Ohr) APS Press, St. Paul, Minnesota, pp. 39-41.

Kapse BM, Pawar VN, Sakhale BK (2009) Postharvest disease management in mango (*Mangiferaindica* L.) cv. Kesar. Acta Hort. 820: 493-503.

Kefialew Y, Ayalew A (2008) Postharvest biological control of anthracnose (*Colletotrichum gloeosporioides*) on mango (*Mangifera indica*). Postharvest Biology and Technology 50: 8–11.

Khan, R. M. & Misra, A. K. 2003. Influence of co-cultivation of marigold, garlic and turmeric on nemic population in guava cropping system. Proc Zonal Conf of ISMPP on Plant Disease of National Importance with Special Reference to Guava Wilt and Mango Malformation. 4-5th Apr, 2003 CISH and IISR, Lucknow, pp., 24-25.

Khan, R. M., Ahmad, I., Kumar, H. K., Singh, A. & Shukla, P. K., 2019. Infestation of root-knot nematode, Meloidogyne enterolobii in newly established/old orchards and nurseries in Madhya Pradesh, Rajasthan and Uttar Pradesh. Ann. Pl. Prot. Sci., 27(1): 170-171.

Khanzada MA, Lodhi M, Shahzad S (2005) Chemical control of Lasiodiplodiatheobromae, the causal agent of mango decline in Sindh. Pak J Bot. 37: 1023-1030.

Kumar, S., Kumar, P., Mishra, P., Kumar, S. and Singh, S., 2017. Management of rust disease (*Ravenalia emblicae* syd.) of aonla (*Emblica officinalis* Gaertn.). Plant Archives 17(1): 292-294.

Lim, T. K. and Manicom, B. Q., 2003. Diseases of guava. In: Diseases of tropical fruit crops. Ed. R. C. Ploetz, CABI publication, UK, pp. 275-290.

Lonsdale, J. H. and Katze J. M. (1993). Chemical control of mango blossom disease and the effect on fruit set and yield. Plant Disease, 77: 558-562.

Masood, A., Shafqat, S. and Asif, S. (2008). Characterization and damage patterns of different bark beetle species associated with mango sudden death syndrome in Punjab, Pakistan. Pak. Entomol., 30: 163–168.

Misra, A. K. & Prakash, O., (1996). Diseases of some minor fruits. In: Advances in Diseases of Fruit Crops in India. Ed. S. J. Singh, Kalyani Publishers, New Delhi, pp. 363-388.

Misra, A. K. (2001). Powdery mildew a serious disease on mango. Journal of Applied Horticulture, 3(1): 63-38.

Pandey, R. S., Bhargava, S. N., Shukla, D. N. and Dwivedi, D. K., (1983). Control of Pestalotia fruit rot of guava by leaf extracts of two medicinal plants. Revista Maxicana de Fitopathologia, 2 (1): 15-16.

Ploetz RC (2009) Management of the most important pre- and postharvest disease. Univ of Florida, TREC, Homestead, Department of Plant Pathology 18905 SW 280th Street, Homestead, FL 33031-3314 USA.

Prakash, O (2004) Diseases and disorders of mango and their management. Diseases of Fruits and Vegetable1: 511-620.

Prakash O, Misra AK, Shukla, PK (2011). Post harvest diseases of mango and their management. Proc Global Conference on Augmenting Production and Utilization of Mango: Biotic and Abiotic Stresses organized by ICAR-CISH at Lucknow during 21-24 June 2011 pp137-144.

Rawal, R. D. and Venugopalan, R. (2011). Disease management in relation to epidemiology of powdery mildew, blossom blight and anthracnose in mango. Souvenir, Global Conference on Augmenting Production and Utilization of Mango: Biotic and Abiotic Stresses, CISH, Lucknow, 21-24 June, pp. 127-130.

Rivera-Vargas LI, Lugo-Noel Y, McGovern RJ, Seijo T, Davis MJ (2006) Occurrence and distribution of Colletotrichumspp. on mango (*Mangifera indica* L.) in Puerto rico and Florida, USA. Pl Pathol J 5: 191-198.

Shukla, P. K. (2020). Mango wilt disease management (Extension folder) ICAR-CISH publication.

Shukla, P. K., Adak, T., Misra, A. K. and Achal Singh (2016a). Appraisal of shoulder browning disease of mango (*Mangifera indica* L.) in subtropical region of India. J. Mycology & Plant Pathology 46(1): 38-46.

Shukla, P. K., Adak T, Gundappa (2017) Anthracnose disease dynamics of mango orchards in relation to humid thermal index under subtropical climatic condition. J Agrometeorology19(1): 56-61.

Shukla, P. K., Bhattacharjee AK, Dixit A (2018a) Efficacy of difenoconazole against shoulder browning disease of mango (*Mangifera indica* L.) and its residue analysis for safety evaluation in mango fruit. IndPhytopathol 71(1): 147-151.

Shukla, P. K., G. Baradevanal, S. Rajan and T. Fatima (2020). MaxEnt Prediction for potential risk of mango wilt caused by Ceratocystisfimbriata Ellis and Halst under different climate change scenario in India. Journal ofPlant Pathologyhttps://doi. org/10. 1007/ s42161-020-00502-9.

Shukla, P. K., Rajan, S. & Fatima, T., (2018c). Root-knot disease of Guava: Identification and Management. Extension folder. ICAR-CISH, Lucknow publication, p. 4.

Shukla, P. K., Savita V, Fatima T, Mishra R, Misra AK, Bajpai A, Gundappa, Muthukumar M (2018b) First report on wilt disease of mango caused byCeratocystisfimbriatain Uttar Pradesh. Indian Phytopathology 71(1): 135-142.

Shukla P. K., Adak, T., Gundappa and Misra A. K. (2016b). Weather based models in predicting powdery mildew of mango for Malihabad mango belt. Journal of Eco-friendly Agriculture, 11(2): 142-148.

Shukla P. K., Misra A. K. and Gundappa (2015). Mango decline and wilt may be a serious threat for future in India. Proceedings of the 36th Annual Conference and National Symposium on "Challenges and management approaches for crop diseases of national importance-Status and prospects" organized by ISMPP, Udaipur held at Madurai, India during 12-15 February 2015, pp. 318-325.

Singh D, Thakur RK, Singh D (2003) Effect of pre harvest sprays of fungicides and calcium nitrate on postharvest rot of kinnow in low temperature storage. Plant Dis Res 18: 9-11.

Singh P (2011) Integrated management of storage anthracnose of mango. J Mycol. and Pl Pathol. 41. 1: 63.

Sinha, P., Chakrabarti, U. and Verma, A. (2001). Critical factors for mango powdery mildew development. Annals of Plant Protection Sciences, 9(2): 264-267.

Sridhar, T. S. and Sinha, P., 1996. Fungal diseases of guava. In: Diseases of fruit crops in India. Ed. S. J. Singh, Kalyani Publishers, New Delhi, pp. 173-182.

Syed NR Mansha N, Khaskheli MA, Khanzada MA, Lodhi AM (2014) Chemical Control of Stem end rot of mango caused by Lasiodiplodiatheobromae. Pak. J Phytopathol. 26(2): 201-206.

Uddin MN, Shefat SHT, Afroz M, Moon NJ (2018). Management of Anthracnose Disease of mango caused by Colletotrichumgloeosporioides: A review. Acta Scientific Agriculture 2. 10: 169-177.

Yang HN, Chen R, Lio R, Hung T (2012) New Botryosphaeriaceae fruit rot of mango in Taiwan: identification and pathogenicity. Botanical Studies 53: 467-478.

Yenjit P, Intanoo W, Chamswarng C, Siripanich J, Intana W (2004) Use of Promising Bacterial strains for controlling anthracnose on Leaf and Fruit of mango caused by Colletotrichum gloeosporioides. Walailak J Sci&Technl (2): 56-69.

7

Harvesting and Post Harvest Protocols for Agro-Advisory in Subtropical Fruit Crops

Akath Singh, Abha Singh, A. K. Trivedi, A. K. Gupta, Karma Beer and Ravi S. C.

Division of Post Harvest Management, ICAR-Central Institute for Subtropical Horticulture, Rehmankhera, Lucknow, Uttar Pradesh

Standard Operating Protocol for Export of Banana

Banana is one of the important fruit grown in India, as far as its export is concerned it is very limited which is below 1%. For export market it should be harvested at 85% maturity or the bunch is ready to harvest in 80-82 days after shooting initiation. Bunches are to be harvested carefully with two people. One person cut the hand and another person put the bunch onto the padding shoulder. Bunches should be slant cut with sickle shaped sharp knife free from any contamination. For export purpose banana fingers should be square, lie close together, hard, clean and free from any infestation and mucilaginous thread of pulp should be 4-5 cm in length. Skin should not be separated from pulp and no brown spot under the skin. De-handing must be done with sharp chisel type de-handing tool, by leaving maximum crown attached to the hand. The crown should be cut evenly; otherwise its outer finger may be detached. Other way to De-handing by using the nylon thread. It is easy method and it allows maximum crown attach with the land with minimum latex flow. The hands should be carefully paced in clean water which has alum 1 gm litre^{-1} or sodium hypochlorite (200 ppm) as surfactant for removing the latex and killing the micro-organism after that fruit must be packed in 20-25 kg capacity crates having foam in bottom and in between hands also. The most effective control of post harvest diseases and control of ripening of banana is treating the hands with 1.0% CISH METWASH an organic formulation for 30-60 seconds after that it should be carefully removed from this organic formulation and keep it in front of high speed fan for drying. The hands should be properly air dried and there should not be any moisture on the fruit surface and cut ends. The care should be taken that hands should not be wiped out with cloth or any other material otherwise organic formulation will be removed. Natural air drying should be given to remove the moisture from skin. The air dried hands are packed in 100 guage polypropylene bags which are lined in 5 ply CFB boxes holding an average weight of 10-13 kg.

The suitable ethylene absorbent (2 sachet) should be placed in corner of the boxes to enhance the shelf life.The air should be removed from the polypropylene bags using the vacuum pump. The polythene lining should be tightly adhered to sking of the fruit without leaving any space to avoid the air build up and water condensation during transit.The packed boxes were stacked and loaded in the pre cooling room to bring down the temperature of the fruit. The relative humidity 85-90% and temperature 13.5 °C of the cooling room should maintain. The banana boxes are directly loaded on to pallets at the pack house, the pallets are kept in reefer containers. The storage temperature 13.0 °C and relative humidity 85-90% should be maintained during transit. As per the protocol developed by ICAR- CISH, Lucknow regarding the ripening of banana. The boxes should be kept in ripening chamber where temperature should be maintained around 25 °C and relative humidity 85-90%. The ethylene gas 150 ppm should be applied for 24 hours. After treating the fruit, it should be kept in ventilated chamber at 13-14 °C temperature with relative humidity 85-90%.

Flowchart for Harvesting and post harvest handling of Banana for Export

Harvesting at 85 % maturity
↓
De handing of the bunches
↓
De-latexing and packing with foam padded plastic crates
↓
Crown trimming and making the cluster
↓
Removal of crown, distorted fingers and dirts
↓
Washing of hands in Alum (1 gm per liter) or Sodium Hypochlorite (200 ppm)
↓
CISH METWASH Treatment (1.0 %)
↓
Air drying
↓
Packaging
↓
Pre cooling (Temp. 13.5 °C, RH 85-90%)
↓
Palletization in cold storage
↓
Loading in reefer containers
↓
Transportation through sea route (Temp. 13.0 °C, RH 85-90%)
↓
Receiving in the destination port
↓
Ripening treatment (Ethylene 150 ppm for 24 hours)
↓
Display in retail unit

CISH – Fruitwash

Sooty mould in mango is a fungal disease which has more than 19 types of fungal spores. It's a secondary infection that's attracted to the secreted honeydew deposited on fruit surface when light rains occur or heavily infected mango by sucking pest they continuously secrets the honey dew and it deposited on the surface of mango fruit. The fungus complex developed in secreted honey dew and in advance stage it turns black and sticky substance adhere on the leaves, stem and peel of mango fruits. Sooty mold doesn't affect the pulp of mango in initial stage but at advance condition it grows on fruit surface rapidly and infect the pulp also. Most of the Mango growing areas of Uttar Pradesh (Middle and Western Uttar Pradesh) are facing the similar problem in their respective field and produce. Because of this infestation its very difficult for the farmers to get remunerative price of the produce. Resulting, farmers are not getting their proper price of mango due to blackening by sooty mould. ICAR-CISH, Lucknow PHM Division identified the above mentioned problem which is prevailing in all mango growing belts of Uttar Pradesh and also analyzes the economic losses because of this malady. To overcome this problem Scientist has developed technology which is eco-friendly wash formulation, which can easily remove black spots lies in the surface of the mango because of sooty mould. Farmers can easily use this wash formulation, by dipping the spotted mangoes in to the solution and kept for 20 minutes. After being removed from the wash formulation, gently rubbing with a smooth cloth can easily remove the spots, transforming the shabby and black spotted mangoes in to shiny one. The post-harvest quality of washed fruit improved significantly as compared to the unwashed fruits (Control) in terms of appearance, shelf life and consumer demand (market price). Use of this formulation can extend the shelf life of the mango up to 8 to 10 days without affecting the fruit quality.

A

B

CISH Anti Ripener I

Once the fruits are harvested, the overall quality of fresh fruits can hardly be improved but it can be maintained through physiological and biochemical interventions. Fresh fruits remain metabolically active even after harvest which makes them highly perishable. Post harvest physiological and biochemical activities of fruit greatly affect produce quality, storability and transportability. Rapid ripening and senescence during handling, retailing and storage greatly impact the external and internal quality of fruits. These changes may be monitored through physiological and biochemical methods. Change in the ethylene level is a remarkable symptom of fruit ripening. For enhancing shelf life of fresh fruits particularly climacteric fruits like mango, it is important to regulate ethylene synthesis. CISH AR I regulates ethylene synthesis pathway and enhances shelf life of fruits.

CISH Anti Ripener II

Huge post harvest losses of fruits are a matter of grave concern for India's horticulture sector. Post harvest physiological changes affect shelf life and the storability of fruits. As a result of stress following the harvest, fruits may synthesize a multitude of chemical compounds, not present in healthy tissue. Many of these compounds detract from quality. Post harvest metabolic changes in fruits are categorized in two parts i. e., anabolic (synthetic) and catabolic (degradative). Ethylene a gaseous plant hormone is considered as responsible for senescence and catabolic changes in fruits. The production of ethylene in climacteric fruits is an autocatalytic process. Initial concentration of ethylene causes an increase in production of ethylene. Hence, to enhance the shelf life of mango fruits it is important that ethylene produced by ripening fruits must be removed immediately. CISH AR II regulates ethylene produced during ripening and enhances shelf life of fruits.

CISH Decomposer

CISH Decomposer
(Consortium of bacterial and fungal cultures)

CISH decomposer is a consortium of bacterial and fungal cultures. Consortia of these microbial cultures are highly active in biodegradation of bio waste. Degradation of agri-horticultural waste is a problem for farmers. This consortium provides a solution for the decomposition of bio waste. Consortia solution in bottles can be sold and transported easily. This should be diluted 4-5 times before use. This solution should be sprinkled on the pile of bio waste, mixed and pile should be covered for 15 days. After 15 days, it decomposes waste. If waste is hard material and pile is thick, 2-3 times sprinkling may be needed.

CISH – Mango Wine Special

Mango (*Mangifera indica* L) is most important fruit crop amongst the subtropical fruit crops. In spite of huge production of mangoes in Uttar Pradesh, mango processing industries are yet lacking in this region. Commercially wine is produced in more than 70 countries. Three of them (France, Italy and Spain) account for almost half of the world's wine production. In spite of tremendous potential and possibilities contribution of India in wine trade is almost negligible. No winery is focusing on wine production from mango. Unavailability of mango wine technology is one of the main constraints. Mango wine prepared with CISH – Mango Wine Special is rich in antioxidants and flavonoids with > 11% alcohol. Moreover, wines are prepared from fresh fruit juice hence; production of fruit wine from a particular fruit becomes a seasonal venture. With this technology wine can be prepared round the year using preserved mango pulp. Hence, raw material will remain available round the year.

Guava Nutri Bite

Guava nutri bite method protocol was developed using beetroot juice fortification. Different combinations of guava pulp and beetroot puree, as well as other components, were employed. The products were evaluated for their physicochemical and sensory qualities, as well as their shelf life under ambient conditions. Other characteristics evaluated included phenolic profile by HPLC, total anthocyanin, TSS, titratable acidity, and ascorbic acid. The physical and chemical changes that happened following the manufacturing of the guava fruit bar were investigated, and a panel of 10 experts used the 9-point hedonic scale to evaluate the product's sensory quality. This fruit bar was stored at room temperature for over 60 days. According to storage studies, T6 (80% guava pulp, 15% beetroot puree, and 5% Others) is the best formulation for fruit bars in terms of physicochemical qualities and organoleptic testing. Additionally, the impact of storage on physicochemical and organoleptic qualities was detected.

Mango Seed Butter from Mango Kernel

The procedure for extracting mango kernel butter has been standerdized. Mango seed butter was extracted from the mango kernel using a Soxhlet equipment and a rotatory evaporator with solvent n-hexane. The working temperatures ranged from 50 to 80°C. The Solvent Extraction technique extracts 14% of butter, with an extraction duration of 5 hours. This butter is highly valued and could serve as a suitable substitute for cocoa or shea butter. Mango kernel butter is a wonderful source of important fatty acids, minerals, and vitamins, and has moisturising characteristics that are useful in skin creams and other products

Low Sugar Aonla candy

Aonla, also known as Indian googberry or phyllanthus emblica,is a fruit native to the Indian subcontinent. its renowed for its high vitamin content and it is used in various culinary and medicinal applications. aonla is often consumed fresh,dried,or processed into products like candy. its valued for its tangy taste and numerous health benefits inclunding boosting immunity,improving degetion,and promoting hairand skin health aonla chakaiya variety cand is a type f candy made from the chakaiya variety of Indian gooseberry (aonla) chakaiya is a specific cultiver known for its superior taste and quality. the candy is typically prepaired by boiling the chakaiya aonla with sugar until it farms a think syrup. which is then allowed to cool and solidity into candy from this variety is particulary prized for its flovor and it popular in many regions for it sweet and tangy taste analysis was found was found in which TSS (Total Soluble Solids) 68:7 Brix and acidity Quantity was 2.24% and vitamin C (99.42) the Conclusion relved that Vitamin C was found in maximum quantity.

Iron and antioxidant Fortified Aonla Candy

Carissa fruits are high in vitamin C, iron, dietary fibre, carbs, lipids, proteins, and micronutrients. This study optimised the sugar syrup levels, dipping time, and drying temperature for developing karonda juice-fortified aonla candy. In this study, we examine the effect of different sugar concentrations (20, 40, and 60 °Brix), time and concentration of dipping aonla segments in karonda juice (30, 60, and 90 minutes), and drying temperature (45°C, 55°C, and 65°C) on the quality of aonla candy. The optimal parameters for producing the intended candy were 60°B sugar syrup and 55°C drying temperature. The corresponding values of responses were: TSS of 55°B, Ascorbic Acid 108. 79 mg/100 g, Titrable Acidity 1.98%, and organoleptic properties such as flavour (8.55), colour (8.95), and texture (8.22). The high TA value may be attributed to the acidic

Bael Aonla Blended Low Sugar RTS

Bael fruit is rich source of riboflavin, thiamine, niacin and minerals. They also contain marmelosin and psorolin which are anti cancerous compounds. Aonla fruits are rich in vitamin C and phenolics compounds with numerous medicinal properties. The powders of mature fallen fruit of bael fruits have restorative and laxative property. Fallen fruit of bael powder provides excellent colour and flavour when they are blended with aonla candy liquid waste. During the aonla candy making the waste in liquid form was utilized by blending with bael powder for preparation of RTS (low sugar immunity booster). The juice, TSS and acidity of RTS are 20-22%, 15°B and 0.25% respectively.

Instant Mango Panna

Mango panna is a refreshing Indian summer drink made from raw mangoes, known for its tangy-sweet taste and cooling properties. This mix is prepared from the freeze-dried mango pulp, infrared dehydrated mint leaf and other ingredients. Freeze-drying process preserves most of the vitamins, minerals, and antioxidants present in raw mangoes, such as vitamin C, vitamin A, and potassium. The developed product having combination of health benefits, traditional significance, and multi-layered taste profile. No preservatives, no added colour is used and the shelf life is six months at ambient temperature. It's particularly popular in regions with hot climates, as it helps to beat the heat stroke. This panna is rich in energy (392 kcal/100 gm), calcium (374.48 mg/kg) and iron (55.77 mg/kg.)

Freeze Dried Ripe Mango Powder

Ripe mango powder was developed through freeze-drying technology to minimize nutritional loss and preserve color, taste, and flavor. Although this process is lengthy and costly, making it inaccessible to most individuals, institutions and entrepreneurs can leverage this technology. The mango powder was incorporated into a baby food mix and underwent organoleptic and proximate analysis for comparison with standardized baby food mixes. Notably, the proximate analysis revealed that the mango powder is a rich source of essential nutrients, including energy (392.0 kcal/100g), fiber (4.03%), iron (34.77 mg/kg), calcium (314.41 mg/kg), and beta-carotene (5.43 mg/100g).

It can be used to develop functional products like smoothies, baked goods, desserts, dressings, or even as a natural flavouring agent for ice creams and yogurts.

Mango Based Baby Food Mix

Three bases were finalized on the basis of organoleptic test of the expert panel in which wheat flour, green gram and almond was same, change was made as variables -rice flake, finger millet and sorghum in same quantity. It can be calculated in the form of ratio as 2:2:1:0.5. All the items are locally available and not costly except small amount of almond which we have taken as a good source of iron, protein and oil. Nutritious weaning mix was developed with combining all the selected three weaning mix bases with developed mango powder in same quantity with same amount of skim milk powder and sugar. The powder improves the taste, color, flavor and appearance of final weaning food mixes. Among these three developed combinations WFM-2 was selected on the basis of organoleptic test and its nutritional composition. It is rich in energy (393 kcal/100 gm), iron (79.75 mg/kg), calcium (372.70 mg/kg), and beta-carotene (0.54 mg/100gm).

Freeze Dried Mango Cubes and Mango- Millet Bar

Dehydrated Mango cubes are preserved for use in different food applications. Total phenolics, ascorbic acid, carotenoids and antioxidant properties of dehydrated cubes were lower in hot air and IR drying technique compared to freeze drying. The millet (Ragi) based mango bar developed along with other nutritious ingredients showed that the bar is highly acceptable, easy to digest, contained high amount of calcium, energy and protein with a shelf life of 60 days. Overall Acceptability was 8. 1 on 9-point hedonic scale

8

Agro-Advisory Services for Fruit Growers of Pune, Maharashtra

Aliza Pradhan, P.B. Taware, G.G. Jena, V.D. Kakade, G.C. Wakchaure, V.N. Salunkhe, S.B. Chavan and K. Sammi Reddy

ICAR-National Institute of Abiotic Stress Management, Baramati-413115

Introduction

Management of abiotic stresses in fruit orchards are essential and key issue for sustaining productivity. Worldwide fruit consumption has increased significantly because of the new recognitions of its broad relevance for health, nutrition and livelihood (Lansky and Newman, 2007). Fruits like pomegranate, grapes, dragon fruit, mango and sapota are being grown in tropical humid to dry as well as arid and semi-arid climates of India, especially in states like Maharashtra, Karnataka, Tamil Nadu, Andhra Pradesh, and Gujarat. Significant cultivars such as Bhagwa of pomegranate, Hapus, Kesar, Langra, Banganpalli, Totapuri of mango and Kallipati, Cricket ball of Sapota are being dominantly grown in these areas. Area under pomegranate was 2.62 lakh ha in 2018-19, producing 30.04 lakh MT at a productivity of 11.58 t ha^{-1} with an export of 80.5 thousand MT during 2019-20 (Maity *et al.,* 2020). The estimated area under grapes, mango, pomegranate and sapota in India is 0.14, 2.29, 0.26, 0.08 million ha, respectively whereas production estimates are reported to be 31.25, 204. 44, 23.15 and 10.03 million MT, respectively (NHB 2019-20 2nd estimate). Besides area of dragon fruit in India is rapidly expanding due to its high nutritional content, antioxidant properties, and market opportunities, covering around 20 thousand ha with annual production of 0.3 million MT. Despite the fact that >15 species of *Hylocereus* have been reported; only 3 to 4 are under cultivation. Among these, white flesh (*H. undatus*) occupied more than half of the area. However, the current premium pricing and consumer demand for pink flesh may shift cultivation to *H. polyrhizus*. Furthermore, species with yellow peel and white flesh (*H. megalanthus*) is now being cultivated. In this context, the issuances of agro-advisory services to farmers are essential for adopting time bound management protocols (Adak *et al.,*

2018a). The importance of this scientific management is significantly linked with the fruit business both in domestic market and foreign export. Therefore, it is of great importance for advisory, training, farmers-scientific interaction and other human resource management programmes to sensitize and develop trained man power sufficiently which contributes to the horticultural growth and development. In this backdrop, fruit orchards and their scientific management is discussed in this chapter with special reference to abiotic and biotic stress management in Pune region of Maharashtra.

Advisory Services to Growers for Abiotic Stress Management in Fruits

Fruit growers are advocated to follow and adapt the advisory issued for time to time for managing various stresses occurring within the fruit orchards. These issuances are highly beneficial to farmers in terms of preparedness and timely implementation of cultivation practices. Pune district of Maharashtra comes under semi-arid region of the Deccan plateau. In general, soils of the district are medium to deep black soils in nature with low water holding capacity. This district experiences a hot semi-arid climate with an average annual rainfall of 744.9 mm, which is very less in comparison to the national average. The fruit orchards of this region encounters several abiotic as well biotic stresses which includes soil stress (low water holding capacity, shallow depth, reduced water availability to plants, soil salinity, etc.), water stress (saline water, waterlogging/hypoxia, deficit water etc.) and atmospheric stress (high temperature during summer, diurnal temperature variationsand high winds) during winter and Monsoon season, erratic rainfall, etc.). So, these stresses need to be addressed significantly for sustainable fruit production in the region. For obtaining optimum fruit growth, maintenance of suitable moisture level is very critical and it is recommended to farmers of mango, sapota, dragon fruit and grape growers to follow the advisory issued for their orchards. Adak *et al.* (2019a) recommended the mango growers for moisture conservation *vis-à-vis* enhancing productivity in mango using drip fertigation. Further, in the semi-arid environment of Pune district, crop losses from diseases and pests are relatively typical. Farmers require repeated prophylactic and curative sprays to manage biotic stressors such as bacterial blight, thrips, fruit borer, and wilt, anthracnose, mildews, leaf and fruit spots, cankers etc. Pest outbreaks such as sucking pest complex, borers, moths, caterpillars and fruit flies needs timely interventions to avoid yield penalty. Agro advisories provide the information and action required by farmers to act, by using sensors to measure things like temperature, humidity, soil moisture, and water circumstances (Nimbalkar, 2019). Precise irrigation, efficient fertilizer use, and nutrient management are necessary to reduce the negative effects of climate. Additionally, it shields the crops from all kinds of infections. Farmers must be aware of the temperature, soil moisture, humidity, and water level to make the most of the agroadvisory. With the use of sensors, field condition reports can be generated, allowing for timely implementation of corrective measures as per the agro-advisory. Using the Agro-Met Advisory Bulletin in accordance with modern technologies can help farmers in doubling their crop productivity and profitability (Kulshrestha *et al.*, 2019). Disease forecasting using various algorithms and software based on long term weather data helps grape growers to follow foliar sprays to control downy mildew, powdery mildew and anthracnose. This agro-advisory section includes key suggestions for the farmer regarding irrigation, nutrient management, and timely management of pests and diseases with appropriate application for Pune district.

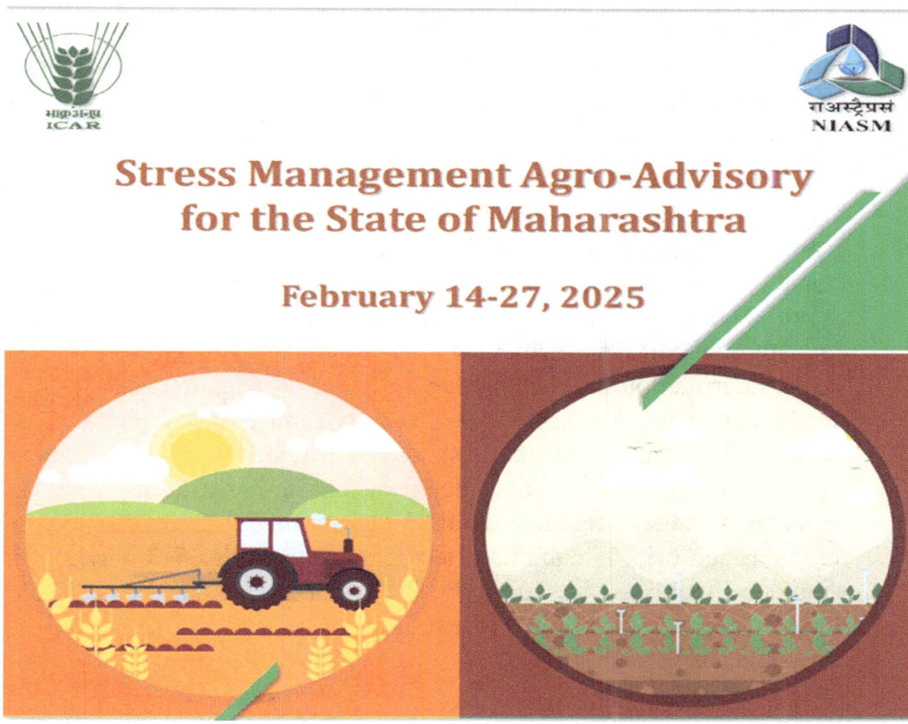

Fig. 1: Agro-advisory services for the benefits of farmers in Maharashtra (published every fortnightly by (*Source*: ICAR-NIASM)

Advisory Services for Managing Water Stress in Fruits

Moisture conservation and irrigation during critical fruit stages is essential to enhance fruit productivity and quality. For producing best quality fruit, assured irrigation is necessary (Patil *et al.*, 2002). Invariably, sudden changes in water availability which may be due to water shortages or high water availability cause fruit fractures as it ripens and reaches maturity, rendering unfit for the fresh market. Therefore, uniform soil moisture level should be maintained using timely irrigation with the use of moisture conservation practices like drip with mulching, use of anti-transpirants like kaolin. Waterlogging should be avoided by ensuring provision for drainage during rainy season to avoid hypoxia in root zone. Further, age of plant, season, location, and management practices all will affect water requirement of fruit trees. Therefore, irrigation schedules for commercial production should consider crop growth stage and irrigation requirements in fruit crops (Chopade *et al.,* 2001; Meshram *et al.,* 2012). In fact, planning of proper irrigation and management includes calculations of crop water requirement based on evapotranspiration data (Meshram *et al.,* 2016). Kumawat *et al.* (2017) suggested using drip fertigation (75 and 60 per cent of water and recommended fertilizers respectively) and other natural resource conservation and management in orchards to enhance resource use efficiency (Adak *et al.,* 2019b). Micro irrigation is advisable for better utilization of water resources and further minimizing losses in mango (Thirupathaiah *et al.,* 2022).

- In grapes, during November month, with the fall in temperature, shoot growth may get hampered which should be managed through sufficient irrigation and nitrogen application during shoot growth stage. Irrigation should be provided during shoot growth stage @ 17,000 to 20,000 ha^{-1} day^{-1} through drip for achieving desired canopy. If shoot growth is vigorous, irrigation should be reduced by half or irrigation can be given on alternate days. During flowering to berry setting stage (approx. 41 to 60 days after pruning), irrigation should be provided @ 5,000 to 7,500 L ha^{-1} day^{-1}through drip for controlled fruit set. Similarly, during veraison to maturity stage, irrigation should be given @ 25,000-30,000 L ha^{-1} day^{-1} through drip. If maximum temperature exceeds 37 °C, irrigation should be increased by 10-15 %. Irrigation should not be withhold during this time else it will affect sugar accumulation and will produce loose bunches.
- In Pomegranate, under water stress areas, planting should be done in rainy season (July-August). However, in water available areas, the spring (February-March) planting is recommended. Drip irrigation of fertigation is usually recommended for the crop. Further, adoption of deficit irrigation and partial root zone drying (PRD) system also saves lots of water and improves WUE (Mohanty *et al.,* 2022) (Fig. 2). In cultivar 'Bhagwa', exogenous use of PGRs like salicylic acid (SA) and napthalene acetic acid (NAA) in conjunction with 80% irrigation levels

can aid in better use of water resources and effectively increasing fruit yield and quality in hot semi-arid climate under shallow murram soils.

Fig. 2: Setup of plot for deficit irrigation and partial root zone drying techniques; (a) Sub main valve (b) Lateral pipe & valve (c) Lateral position in case of DI (d) Lateral position in case of PRD (e) 2 drippers per plant (f) A single dripper with discharge capacity 4 L h^{-1}

- In case of mango, irrigation should be given after 20-30% of flower initiation, so that good flowering can be achieved. Further, in case of new plantations, care should be taken to avoid waterlogging in case of high rainfall areas and proper drainage shall be ensured. Adequate soil moisture should be maintained in case of dry spell to ensure vegetative growth during the initial years and higher survival in the field. Good flowering and fruit setting as well as growth can be achieved through use of NAA 4.5 L @0.5 ml L-1 during fruit setting and during initial fruit growth (pea size) along with SA @300 ppm.
- In Dragon fruit, under heavy soils, the performance of dragon fruit is hindered by water logging. Water logging invites stem rotting and other infections due to several saprophytes. Hence, proper drainage is an important aspect to avoid water logging. The water logging in dragon fruit orchards can be avoided using raised bed planting. The dimensions of raised bed can be customized as per the site *viz*., (5-6) × (0.5-1) feet for light soils and, (5-6) × (2-3) feet for heavy soils, respectively. During heavy rainfall events, provision should be made to remove excess water from fields.

Advisory Services for Managing Soil Stress in Fruits

Management of soil stresses is highly important for ensuring fruitful production. In this context, advisories were issued to farmers for managing soil stresses. As arid soils have low levels of organic matter and nutrients, effective management of soil health and fertilizer is necessary to achieve good yield (Kumar *et al.*, 2018). Nutrient deficiencies are major cause of concern for low productivity. In fact there were huge gaps in yield of fruits as a function of soil properties (Adak *et al.*, 2020b). Therefore, scientific technologies should be adapted to enhance soil and orchard performances.

- Quality of the fruits is decided by the availability of the nutrients. Hence, many times growers are advocated for foliar application of Potassium @ 0.3% (potassium metalosate) alone and with combination of Boron applications (0.15 and 0.3% H_3BO_3) for improvement in fruit quality (Tehranifar and Tabar, 2009).
- Growers were advocated to apply recommended manure and fertiliser dosage in the month of January, June, and September for non-bearing trees (1-3 years). The annual requirements for a fully developed tree (five years or older) are 40 kg FYM, 625 g nitrogen, 250 g phosphorus, and 250 g potash.

- Even, farmers were told to follow foliage zinc nutrition technology (1 % $ZnSO_4$ application) in guava improved nutrient status, yield and fruit quality (Adak *et al.*, 2020a).
- In case of low organic matter FYM, poultry manure, vermicompost and microbial inoculation need to be applied.
- Growers must employ best management practices i. e., effective irrigation, proper canopy management, cover or intercropping, over-tree sprinklers, shade netting, fruit bagging, suppressants (Kaolin or calcium carbonate), and chemical protectants (Lal, 2017).

Different micronutrient and their deficiency symptoms along with management practices are given in Table 1.

Table 1: Agro-advisory to farmers for micronutrient deficiencies and its management

Micronutrient	Deficiency symptoms	Management practices
Boron	Affects pollen development. Premature shedding of male flowers which affects fruit setting. Causes fruit cracking	Spraying of boric acid @0.05-1% (one during pre-blooming and two during post blooming period)
Copper	Poor shoot growth Small fruits of inferior quality	Copper sulphate (0.2%) spray before flowering
Iron	Yellowing of leaves	Foliar application of Fe chelates (0.3-0.5%)
Manganese	Inter veinal chlorosis	Application of manganese sulphate (0.4-0.6%)
Zinc	Poor flowering and fruit set Small size fruits	Foliar application of zinc sulphate (0.3-0.5%) before bud opening. Application of EDTA-Zn (2.5-7.5g per plant) to soil.

Celebration of World Soil Day (*Source*: ICAR-NIASM)

- Dragon fruit can be grown successfully in diverse soils, including sandy, sandy loam, sandy clay loam, clay dominating (black, red). Establishment of an orchard at ICAR-NIASM demonstrated that it can be successfully cultivated in rocky, barren, degraded lands with poor soil formation and low water holding capacity (Nangare *et al.,* 2020; Wakchaure *et al.,* 2021). It can be grown successfully in soil with a pH of 5.5-8. 0, but prefers organic matter in soil, slightly acidic pH and can tolerate some salts. Root studies conducted at ICAR-NIASM, revealed plant roots hardly reach up to 1.5 m either horizontally or vertically in soil profile. Most of root mass (50%) is confined to 30-60 cm and thus, soil depth may not be constrain for cultivation. In dragon fruit, it has been observed due to increase in air as well as soil temperature results in to increase in sunburn and stem/soft rot. Therefore, mulching with sugarcane trash results in reduction of these issues through saving of water and regularization of soil temperature (Fig. 3 to 4). Further, NPK bio consortium or humic acid @ 1 L per acre should be given through drip to make nutrients available to plants, Fertigation with 19:19:19 grade is also recommended.

Fig. 3: Sugarcane trash mulching in dragon fruit (*Source*: Farmers' field)

Fig. 4: Visit to dragon fruit orchard of (*Source*: ICAR-NIASM)

- The grape vineyard should be established on pure, true to type rooted cuttings of suitable rootstocks for the specific agro-climatic region (Table 2). Best time of planting rootstocks in field is February to March. Open trenches of 60-75 cm depth and 60-75cm width in north-south direction with spacing of 2.75- 3.00 m should be made at least a month before planting and should be filled with native soil, FYM, fertilizers and crop debris.

Table 2: List of rootstocks recommended for different soil and water conditions.

S. N	Situation/ Problem	Rootstock
1	Water shortage	1103P, 140Ru, 110R, 420A, SO4, Dogridge, 99R.
2	Soil EC more than 2mmhos/com and water EC more than 1mmhos/cm	Ramsey, 140Ru, 99R, 110R.
3	Soil ESP more than 15 per cent and / or water SAR more than 8.	140Ru, Ramsey, 110 R, 1103P
4	Free calcium content of soil >12%.	Fercal, 140Ru, SO4, 420 A.
5	Chloride content of water > 4 meq/lt	Ramsey, Dogridge, 140Ru, Teleki 5C, 110 R, 99 R.
6	Poor vigor of the variety without any soil/ water problem.	Dogridge, St. George, SO4, 140Ru.

- Soil application of FYM @ 25 metric ton and SSP @ 200 Kg ha^{-1} should be given 8-10 days before pruning. Other nutrient requirements can be achieved through fertigation. During shoot growth stage, urea @ 50-75 Kg ha^{-1} in four splits should be given, based on soil test report. At 5-leaf to pre-bloom stage, Zinc sulphate and Ferrous sulphate should be applied @ 30 Kg ha^{-1} each. Boron application @ 2.5 Kg ha^{-1} should be carried out only if soil test value indicates low levels. To facilitate nutrition management in orchard, petiole nutrient testing to be carried out at 70% cap-fall stage (approx. 50 days after pruning). The petiole samples of leaf opposite to the bunch should be removed for sampling (Fig. 5 to 6).
- A technology for establishment of Sapota orchard on superficially subdued basalt igneous rocky area has been developed and validated at ICAR-NIASM, Baramati. In this technology, pits of 1x1x1m dimensions are prepared with the help of excavator machine. As the site was rocky, micro-blasting was done for cracking the bottom rock so that depth was further enhanced to 1 m. The pits are filled with soil mixture comprising native murrum soil with black soil in 1:1 proportion along with 20 kg farm yard manure (FYM) and 500 g single super phosphate (SSP) fertilizer. The Sapota saplings (Kalipatti) are planted in the pits followed by watering. The standard canopy management practices are followed to achieve satisfactory development and fruiting of the plants during the subsequent years. This technology has resulted in to getting 11.43 ton/ha of Sapota yields. Besides, to achieve good growth of fruits, alternate fertigation with 19:19:19 and 0:52:34 fertilizer grades along with NPK bioconsortium @1 L per acre is recommended.

NIASM Joins hand with Grape Farmers Association of Maharashtra to Address Abiotic Stress Issues

ICAR-NIASM, Baramati hosted the 2nd research advisory committee (RAC) meeting of Maharashtra Rajya Draksha Bagaitdar Sangh (MRDBS), Pune on 10th November 2022. In the forenoon, all the members of RAC and the board of directors of MRDBS witnessed efforts and innovation of farmers in making the grape orchard climate proof by plastic-cover. The initiatives of grape growers of Nirgudi in Phaltan Tehsil of Satara, Maharashtra provided first-hand information on the researchable issues which were discussed in detail in the afternoon session, the RAC meeting that was started with a traditional welcome by Dr. Jagadish Rane, Director, ICAR-NIASM, Baramati. He briefed the RAC members and the board of directors of MRDBS about the institute's achievements and ongoing research activities. Further, he highlighted the role of ICAR-NIASM in abiotic stress management and how it could be extended its research facilities to address the abiotic stress management in Grapes. The RAC meeting was chaired by Dr. G.S. Prakash, formers Principal Scientist, ICAR-IIHR, Bengaluru and a very popular knowledge partner for Grape Growers Association. All the members of RAC discussed the research activities to be taken up during this cropping season pertaining to berry cracking, nutrient uptake pattern in Crimson grafted on Dogridge and Polson rootstock, berry drop and rot, standardization of protected structures for grapes, and other research issues raised by the member. The RAC meeting was attended by Dr. S.D. Sawant, VC, BSKKV, Dapoli; Dr. R.G. Somkuwar, Director, NRC Grapes; Dr. R.A. Marathe, Director, NRC Pomegranate; Dr. Sujata Tetali, ARI, Pune; Dr. Vijay Chole; Dr. J.M. Khilari and Dr. Jagadish Rane, Director, ICAR-NIASM, Baramati. The suggestion emerged from deliberation are expected to result in collaborative research and opportunities for young research scholars who can contribute enhancement of income of farmers through their efforts. Detailed program of research will be submitted to the MRDBS and also funding agencies to achieve these objectives. The RAC meeting was coordinated by Dr. P.S. Khapte, ICAR-NIASM, Baramati, Mr. Sunil Pawar, and Mr. J.N. Kalbhor from MRDBS, Pune.

Fig. 5: Extension activities for grape growers of Maharashtra (*Source*: ICAR-NIASM)

Fig. 6: Progressive farmers visit in grape orchards of (*Source*: ICAR-NIASM)

Advisory Services for Managing Biotic Stress in Fruits

Horticulture has huge potential to improve socio-economic status in India (Singh *et al.*, 2021) and therefore, growers benefit should be the aim of scientific approaches. Biotic stress management is the key for successful fruit production. Awareness, sensitization, hands on training and other farmer-scientist interaction programmes ensures the benefits of farmers' community.

(Adak *et al.*, 2018b) explained scientist-farmers interaction on regular basis for diffusing horticultural technologies to small, marginal and progressive farmers are the important parameters for improving income. Therefore, biotic stress management is obvious and should be taken into consideration for commercial fruit production (Fig. 7 to 8).

- A large number of diseases and insects damages pomegranates. In India, bacterial blight and wilt pose a serious threat to pomegranate. Pin hole borer (*Xyleborus perforans*), Thrips (*Rhipiphorothrips cruentatus*), aphids (*Aphis punicae*), mealy bug (*Ferissia virgata*), and anar butterfly (*Deudorix iscocrates*) are the major pests. Adults and nymphs of pests suck the cell sap from shoots, leaves, and fruits. Farmers were advised to follow management protocols for enhancing productivity (Table 3).

Table 3: Agroadvisory for management of pests in pomegranate for farmers' benefits

Sl No	Insect/Pest	Symptoms	Pesticide/Insecticide
1	Mites (*Tenuipalpus punicae*)	Brown patches on the under surface of affected leaves	Spraying of propargite 57% @ 1. 0 ml/l, spiromesifen 240 SC @ 0.5 ml/l, or dicofol 18.5 EC @ 2.0 ml/l
2	Thrips (*Rhipiphorothrips cruentatus*)	Affected leaves curl upwards and downwards	Imidacloprid @0. 4 ml/l or Acephate 75 SP @ 1ml/l on
3	Fruit borer (*Deudorix isocrates*)	Brackish to brown colour holes on the surface	Spray neem seed extract @50g/l during flowering and spraying of deltamethrin 2.8 EC @ 1.5 ml/l or malathion 50EC @ 2ml/l or azadirachtin 1500 ppm @ 3.0 ml/l at 10-15 days intervals up to harvesting
4	Aphids (*Aphis punicae*)	Sticky substances on the surface	Spray imidacloprid 0.3 ml/l, thiamethoxam @ 0.4ml/l or dimethoate 1.5 ml/l
5	Fruit sucking moth (*Eduocima maternal*)	Punctured holes on the surface	Keeping poison baits (95% jaggery + 5% malathion) in the field
6	Termite	Tunnel in the trunk	Spray malathion (1.5 ml/l) Drenching soil with chloropyriphos (2.0 ml/l)

Sl No	Insect/Pest	Symptoms	Pesticide/Insecticide
7	Bacterial blight (*Xanthomonas axonopodis*)	Water-soaked spots on underneath of leaves	Heavy pruning in case of severe infection. Spraying of Bordeaux mixture altered with streptocycline (0.5g/l)/bronopol @ 0.5g/l mixed with copper oxychloride (2-2.5g/l).
8	Wilt (*Ceratocystis fimbriata*)	Foliage turns yellow and plant starts dying	Drenching soil with chloropyriphos 20EC (2.5 to 4.0 ml/l) + carbendazim 50WP (2.0g/l) or propiconazole 25EC (2ml/l) using 5-8 litre solution per plant
9	Nematode (*Meloidogyne sp.*)	Leaves of infested plants turn pale green and yellowish colour with stunted plant growth and resulting in wilting and drying of the plant	Application of Neem cake @ 1-2 kg per plant
10	Fungal spots and rots	Dark black to brownish Black of irregular shape and size on fruits	Spraying of mancozeb (0.2%) + carbendazim (0.2%)

- In dragon fruit, being new introduction to the country, very few diseases and pests were observed in the initial years of introduction. However, in recent years many diseases, pests and disorders are being observed in the dragon fruit. However, at present growers are less aware with diseases, pests and their management in dragon fruit. During a survey of farmers' fields, diseases such as stem canker, anthracnose, soft rot, brown spot, and stem blight were frequently observed. These diseases are exacerbated by high soil moisture, atmospheric humidity, and temperature. Viral disease ('X' virus) is now being observed. Use of disease-free planting material, sanitization of pruning tools etc. is required for the establishment of a healthy orchard. Pests like fruit flies, aphids, termites and birds causes economic yield losses. Use of pheromone traps, fruit covers, and other prophylactic sprays will aid in the prevention of fruit fly and bird attacks. However, this crop has shown good tolerance to many pests and diseases. Still, preventive measures such as drainage to avoid waterlogging and optimal shade, orchard sanitation, and monitoring and use of preventive chemicals, at regular intervals and timely nutrient management are the key factors to keep plant healthy and pest-disease free.
- In Mango, proper care of pest and diseases shall be taken during flowering and fruiting. Major pests include thrips, aphids, fruit fly,

gall midge and diseases like anthracnose, powdery mildew are being observed. Spraying of Carbendazim 12% + Mancozeb 63% WP (@ 0.2%) and CoC (50% WG) @ 0.24% at the interval of 10 days shall be taken to control anthracnose and powdery mildew. For control of sucking pests, to control at initial level Azadirachtin 1% @ 2-3 ml per liter shall be sprayed. Also insecticides imidaclopride 17.8 SL @ 0.3/ litre or imidachlopride+cypermethrin (1-2 ml per litre), or Spinosad 45% SC 05-1ml per litre may also be used as per pest populations. shall be sprayed. Whereas, in Sapota leaf webber (chiku moth) and hairy caterpillars during the rainy season are the common pests on new flush and flower buds. Spray two rounds of carbaryl 0.1 per cent, chlorpyriphos 0.05 per cent or quinalphos 0.05 per cent at 20 days interval after new shoot formation. In case of leaf spot disease, plant shows symptoms like numerous small pinkish to redish brown spots on the leaves. It is advised to spray Mancozeb @ 0.25% or Copper oxychloride 0.2% to control.

- In grapes, bordeaux mixture 1% or Copper oxy chloride @ 2.5g L^{-1} should be sprayed along with wettable sulphur 2g L^{-1} immediately after pruning to reduce pathogen inoculum load in orchard. Further, flea beetle infestation may increase during early shoot growth. For the management of flea beetle, Fipronil 80 WG @ 0.06 g L^{-1} (do not use during and after flowering stages) should be sprayed. At 15 days interval, washing plant with entomopathogenic fungi like Metarhizium, Beauveria and Lecanicillium will be useful for controlling mealybugs and ants. Mandatory spray schedule should be followed for downy mildew management even if prevailing conditions are not conducive for the disease. Spraying of Buprofezin 25 SC @ 1.25 ml L^{-1} (PHI 65 days) or plant wash with Trisiloxane polyether surfactant @ 0.3 ml L^{-1} followed by plain water wash is advised for mealybug and mite infestation. Sparying of sulphur 80 WDG @ 1.5-2.0 g L^{-1} or Abamectin 1.9 EC @ 0.75 ml L^{-1} (PHI 30 days) is recommended for only mite infestation. However, to avoid pesticide residue issues, recommendations published by National Research Centre for Grape, Pune for pest and disease management should be followed.

Fig. 7: Farmer-scientist interaction on pomegranate wilt (*Source*: ICAR-NIASM)

Fig. 8: Awareness programme on Bacterial blight and wilt in pomegranate organised (*Source*: ICAR-NIASM)

General Recommendations for Fruit Growers

Vineyard Establishment and Canopy Management in Grapes

- Allow all the shoots growing after planting till 50-60 days. Thereafter, retain three vigorously growing shoots and tie them straight with the support of bamboo stick (Fig. 9).
- The best season for *in situ* grafting is the month of September and October. Follow wedge grafting on two healthy rootstock shoots so as to get cent percent successful grafting. Select good quality, healthy and 8-10 mm thick mature scion sticks for better stand.

- Do not disturb graft union and irrigate to avoid desiccation. Remove rootstock suckers time to time to avoid competition for water and nutrients.
- As the sprouted scion grows the plastic tape at graft joint should be loosened to avoid girdling and removed subsequently as the graft union is achieved.

Fig. 9: Scion bud sprouting after grafting and subsequent growth (*Photo Courtsey*: P B Taware)

- Recut on the scion by leaving 2-3 eye buds above graft union is required during preceding February month to maintain uniform growth in whole orchard.
- The sprouted shoots are trained to develop main stems, primary arms and cordons of the grapevines.
- Grape originally being temperate fruit crop, at tropical regions double pruning single cropping pattern need to be followed.
- The pruning operation should be staggered based on local climatic conditions and market status.
- Back pruning should be carried out during March to April period for development of fruiting units in the form of developing canes and food storage. Only one or two eye buds should be retained on the cordons during pruning (Fig. 10).
- If the shoot vigour is more, sub-cane technique should be followed for assured fruitfulness.

- Forward pruning can be staggered from August to November months considering local climatic conditions. Efforts should be made to adjust pruning time so as to avoid rainy days coinciding flowering and maturity stages (Fig. 11).
- It is advised to carry out sample cane check for fruitfulness analysis to decide pruning zone. As a thumb rule pruning should be carried out after single eye bud in case of sub cane and immediately after shortened internode between 6-10 buds in case of straight cane.
- Canopy management standards should be followed after both the pruning during production cycle as mentioned in Table 4.

Table 4: Standards for canopy management in grape during annual production cycle

Particulars	**Canopy Standards**
Spacing	2.5-3. 0 m x 1.5-1.8 m
Area/vine	3.75 to 5.4 sq. m
No. of canes/sq. m (post back pruning)	7 - 8
No. of shoots/sq. m (post forward pruning)	8 - 10
No. of bunches/sq. m	8 - 10
No. of bunches/vine	30 - 60
Bunch wt. (g)	300 - 400
Yield/vine	12 Kg to 18 Kg
Yield/ha	30.0 MT

Fig. 10: Back pruning, bud sprouting, sub-cane pinching and subsequent canopy development in grape (*Photo Courtesy*: P B Taware)

Fig. 11: Forward pruning in grape and pasting with bud break solution (*Photo Courtesy*: P B Taware)

- Canopy management should be looked after carefully. Retain 8-10 shoots along with 8-10 bunches per square meter area. Foliar application of Gibberellic acid 10 ppm at pre-bloom stage is required in Thomson seedless for rachis elongation. At 50% flowering stage, bunch treatment with 20 ppm GA3 is recommended to reduce fruit set and for rachis elongation.
- During December, manage canopy for adequate sunlight and ventilation within the canopy to avoid risk of inflorescence necrosis. Minimum temperature will be low in some grape growing areas resulting into slow growth. Foliar spray of Sulphate of potash @ 2-3 g L^{-1} during pre-bloom stage and fertigation with SOP @ 40-60 kg ha^{-1} will help in stress alleviation.
- In the month of January, to avoid cold stress due to cool air drift, cover orchard sides vertically with clothes. Foliar spray of Sulphate of potash @ 2-3g L^{-1} during pre-bloom stage and fertigation with SOP @ 40-60 Kg ha^{-1} will help in stress alleviation.
- At the berry size is from 2-4 mm and 5-7 mm, GA3 application @ 40-50ppm is required in Thomson seedless and Sharad seedless. Spray Calcium Chloride @ 2 g or Ca chelate 0.5 g L^{-1} immediately after GA application. Complete berry thinning operations meanwhile to retain 80-90 berries per bunch.
- In the month of February, arrange the grapevine shoots on trellises so as to create canopy cover for bunches to protect from sunburn injury. In case of sparse canopy, use bunch cover or shading net options to avoid sunburn (Fig. 12).

Fig. 12: Sunburn in open canopy and its management through shade net covering in grapes

- Grape is a non-climacteric fruit. Hence, it should be harvested at perfect maturity. Minimum TSS of 16 °Brix and sugar acid ratio of 20 (AGMARK standard) should be ascertained during harvesting.
- While harvesting, packaging and transport, careful handling of grape bunch is must to avoid bruising of berries and subsequent drying and decay damages.

Bahar Treatment in Pomegranate

It was known to all that the pomegranate (*Punica granatum* L.) can be cultivated in different agroclimatic conditions (Jalikop, 2007) and due to its wide range of adaptability, hardiness, low maintenance requirements, and excellent returns (Singh *et al.*, 2007). The most significant cultivar is "Bhagwa," Maharashtra (Pal *et al.*, 2017) and India is the largest producer of pomegranates in the world (Meshram *et al.*, 2019). In pomegranate, *bahar* treatment is being followed to regulate the crop as it can flower thrice a year. There are three flowering seasons in dry and semiarid climates: February to March (Ambe bahar), July to August (Mrigbahar), and September to October (Hasta bahar) (Pal *et al.*, 2014). The choice of flowering or fruiting depends on irrigation water availability, market demand, and the prevalence of pests and diseases in a particular area.

- Ambe bahar fruit matures between June and August and needs reliable watering to be ready for harvest. Because fruit growth occurs during the dry months, this bahar fruit has poor size and colour, and its yield is very low. There are more instances of pests like mites. As a result, ambe bahar is not considered a hot, dry region.
- December through February are the months during which the mrig bahar crop is harvested. The fruits grow during the wet season and finish maturing in the winter. Due to the scarce water supply and environmental conditions, this bahar is popular in arid and semi-arid regions. However,

fruit cracking is more due to fluctuations in day and night temperature. ICAR-NRC on pomegranate doesn't recommend farmers to adopt Mrig bahar in orchards having severe bacterial blight problem (https:// nrcpomegranate. icar. gov. in).

- In the months of March and April, fruits of hasta bahar are harvested. They feature very lovely dark-coloured arils on their rind. Fruits cost a high price because there is a restricted supply during this season. The absence of irrigation during this time corresponds with the rainy season, preventing the development of optimal water stress. Poor flowering and fruit set result from this lowers output. As a result, to achieve a high yield with sustainable production in pomegranates, adequate canopy management and flower regulation are necessary. Sometimes growth regulators were also advised to spray for fruit settings and harvesting good yield (Anawal *et al.,* 2016).
- Bahar treatment involves stopping irrigation one to two months before the bahar. After withholding watering for 40-45 days, the trees receive mild pruning. Following pruning, fertilizer applications are made at the required rates followed by irrigation which results in abundant fruiting and flowers. Three flushes of bisexual, intermediate, and male flowers are produced by pomegranates on both new and old growth. Four to five months after blossoming, the fruits are ready for harvesting.

Fruit Cracking and Sunburning

To get rid of fruit cracking and sunburning, fruit growers were advised to take care of their orchards.

- Cracking is brought on by nutrient deficiencies and an unbalanced utilisation of nutrients. Fruit cracking in young orchards is caused by boron shortage, whereas in older orchards, moisture imbalance and temperature fluctuations are key contributing factors.
- Fruits typically split when it rains or when irrigation is applied after a protracted dry spell. Farmers were suggested to apply sufficient and consistent irrigation during the fruiting season to prevent cracking.
- The amount of cracking can be decreased by spraying gibberellic acid at a concentration of 20 ppm, calcium chloride at a concentration of 2 ppm, and boron at a concentration of 0.2 ppm during fruit maturity.
- Cover fruits with bags during hot, dry weather to retain fruits. Damage due to sunburning can be decreased by applying a 5% kaolin spray during the months of May to July. On pomegranate fruit, shading with a 35% shade net can assist to lessen sunburn.

- Fruits should be bagged with cover bags to reduce the risk of sunburn. White coloured bags reflect sunlight more effectively and prevent attack of fruit fly.

References

Adak, T., Kumar, K., Pandey, G. andSingh, V. K. (2020a). Response of $ZnSO_4$ through foliar nutrition on nutrient status, yield performance and fruit quality of guava cv. Shewta. Current Advances in Agricultural Sciences 12(2): 93-97.

Adak, T., Pandey, G., Pandey, D., Garg, N., Ram Kumar, Singh, S. R., Chandra, S. and Kumar, A. (2018b). Scientist-Farmers interaction for enhancing farmers' income through nutrient management, crop diversification, value addition and other horticultural technologies. In: National Conference on Strategies & Challenges on Doubling Farmers' Income through Horticultural Technologies in Subtropics, held during June21-22, 2018 at CISH, Lucknow, pp. 115.

Adak, T., Pandey, G. and Babu, N. (2020b). Yield Gap Analysis in Fruit Orchard as a Function of Soil Physico-Biological and Chemical Indicators. Journal of Agricultural Physics, 20(1): 1-14.

Adak, T., Pandey, G. and Rajan, S. (2019b). Advanced knowledge on soil and water conservation measures in fruit orchards. Journal of Agricultural Physics 19(1): 35-45.

Adak, T., Shukla, P. K., Singh, A. K., Gundappa, Chandra, S. (2018a). A new approach to reach farmers through agroadvisory services for better orchard management. Kahaar, 5(3): 31-36.

Adak, T., Singh, V. K., Kumar, K. and Pandey, G. (2019a). Moisture Conservation and Productivity in Mango cv. Langra using Drip Fertigation. In: Soil and Water Conservation Bulletin, Mishra, P. K., Sharma, N. K., Kumar, A., Mandal, D., Kaushal, R., Kumar, G., Roy, T. and Dogra, P. (Eds), Indian Association of Soil and Water Conservationists, Dehradun, Uttarakhand, 4: 77-80.

Adsule, P. G., Yadav, D. S., Upadhyay, A., Satisha, J. and Sharma, A. K. (2013). Good Agricultural Practices for Production of Quality Table Grapes. Technical Bullettin pp. 64. Published by Director, NRCG, Pune.

Anawal, V. V., P, Narayanaswamy and Ekabote, S. D. (2016). Effects of plant growth regulators on fruit set and yield of pomegranate cv. Bhagwa. International Journal of Current Research 8(04): 29008-29010.

Chopade, S. O., Gorantiwar, S. D., Pampattiwar, P. S. and Supe, V. S. (2001). Response of Pomegranate to drip, bubbler, and surface drip irrigation methods. Adv. Hort. For. 8: 53-59.

Jalikop, S. H. (2007). Linked dominant alleles of inter-locus interaction results in a major shift in pomegranate fruit acidity of 'Ganesh' and 'Kabul Yellow'. Euphytica 158: 201-207.

Kulshrestha, M., Nitin, P., Dhabale, S. and Kumar, V. (2019). Exposure, Perception and Advantages about Weather based Agro-advisory Services by Selected Farmers of Anand District, India. International Journal of Current Microbiology and Applied Sciences 8. https://doi. org/10. 20546/ijcmas. 2019. 805. 224.

Kumar, R., Meena, R., Sharma B. D. and Saroj P. L. (2018). Production technology of pomegranate in arid region. CIAH/Tech. /Bull. No. 65, ICAR-Central Institutefor Arid Horticulture, Bikaner, Rajasthan, India.

Kumawat, K. L., Sarolia, D. K., Kaushik, R. A. and Jodha, A. S. (2017). Effect of irrigation and fertigation scheduling on growth, flowering, yield and economics of guava cv. Lalit under ultra high density planting system. Indian Journal of Horticulture 74(3): 362-368.

Lal, N. and Sahu, N. (2017). Management Strategies of sun burn in fruit crops-A review. Int. J. Curr. Microbiol. App. Sci. 6(6): 1126-1138.

Lansky, E. P. and Newman, R. A. (2007). Punica granatum (pomegranate) and its potentialfor prevention and treatment of inflammation and cancer. Journal of Ethnopharmacology 109: 177-206.

Maity, A., Babu, K. D., Singh, N. V., Patil, P. G. and Gaikwad, N. (2020). ICARNRCP Annual Report 2020, ICAR-National Research Centre on Pomegranate, Solapur- 413 255,Maharashtra.

Meshram, D. T., et al. (2019). "Crop Coefficient and Evapotranspiration of Pomegranate (Punica granatum L.) for Western Part of Maharashtra, India". Acta Scientific Agriculture 3 (7): 218-223.

Meshram, D. T., Gorantiwar, S. D., Mittal, H. K., Singh, N. V. and Lohkare, A. S. (2012). Water Requirement of Pomegranate (*Punica granatum* L.) plants upto five year age. J. App. Hort. 14(1):47–51.

Meshram, D. T., Singh, N. V. and Pal, R. K. (2016). Improvement of water use efficiency in pomegranate (*Punica granatum*) cv. Bhagwa under micro-irrigation system. Indian Journal of Agricultural Sciences 86(2): 192-6.

Mohanty, S., Panigrahi, P., Samant, D., Raychaudhuri, M. and Ambast, S. K. (2022). Effect of partial rootzone drying on growth and yield of mango plants. Indian Journal of Soil Conservation 50(1): 66-71.

Nangare, D. D., Taware, P. B., Singh, Y., Kumar, P. S., Bal, S. K., Ali, S. and Pathak, H. (2020) Dragon Fruit: A Potential Crop for Abiotic Stressed Areas. Technical Bulletin No. 47. ICAR-National Institute of Abiotic Stress Management, Baramati, Pune, Maharashtra, India, pp: 24.

Nimbalkar, A. and Patil, Y. (2019). Using Wireless Sensor Network for Acquiring Environmental Parameters for Designing An Early Warning System For Pomegranate Farmers: A Review. i-Manager's Journal on Wireless Communication Networks 7(4), 35.

Pal, R. K., Babu, D., Singh, N. V., Maity, A. and Gaikwad, N. (2014). Pomegranate research in India-Status and future challenges. Progressive Horticulture 46(2): 184-201.

Pal, R. K., Sharma, J., Babu, D., Singh, N. V. and Gaikwad, N. (2017). Pomegranate Revolution in India- A success story of ICAR-NRC on Pomegranate Solapur. 14p.

Patil, A. V., Karale, A. R. and Bose, T. K. (2002). Pomegranate. In: Bose, T. K., Som, M. G. (Eds.), Fruits: Tropical and Subtropical. Naya Prakash, Kolkata, India, pp. 225-262.

Singh, A. K., Sanakaran, M. and Murthy, B. N. S. (2021). Horticulture for food, nutritional and socio-economic security in India. Indian Horticulture 9-10: 1-9.

Singh, D. B., Kingsly, A. R. P and Jain, R. K. (2007). Studies on separation techniques of pomegranate arils and their effect on quality of anar dana. Journal of Food Engineering 79: 671-674.

Tehranifar, A. and Tabar, S. M. (2009). Foliar application of potassium and boron during pomegranate (Punica granatum) fruit development can improve fruit quality. Horticulture, Environment and Biotechnology 50(3): 191-196.

Thirupathaiah, G., Bhagwan, A., Kiran Kumar, A., Avil Kumar, K. and Vijaya, D. (2022). Impact of irrigation and fertigation alongwith mulching on fruit yield, water use efficiency, nutrient use efficiency and economy of mango. Indian Journal of Soil Conservation 50(1): 72-84.

Wakchaure, G. C., Satish Kumar, Meena, K. K., Rane, J. and Pathak, H. (2021). Dragon Fruit Cultivation in India: Scope, Constraints and Policy Issues. Technical Bulletin No. 27. ICAR–National Institute of Abiotic Stress Management, Baramati, Pune, Maharashtra, India, p. 47+vi.

9

Farmer's Advisory for Sustaining Productivity in Small Cardamom Black Pepper and Turmeric

C. N. Biju, A. Jeevalatha, M. Alagupalamuthirsolai and M. S. Shivakumar

ICAR-Indian Institute of Spices Research, Kozhikode-673 012, Kerala, India

Introduction

Spices have been considered significant in the culinary art from time immemorial. India, "The Homeland of Spices" is one among the largest producers, consumers and exporters of spices as well as spice products. Spices comprise different plant components such as floral parts or fruits or berries or seeds or rhizomes or roots or leaves or kernel or aril or bark or bulbs. The recorded history of the use of spices dates back to 5000 years or so from the Egyptian or at the most from the Indus valley civilization. Spices are used for flavouring, seasoning and imparting aroma in various cuisines. Besides these inherent qualities, some of them are known to be fungistatic, antimicrobial or antibiotic. They are also processed into numerous 'value added' products such as oleoresins, essential oils, curry powder etc. Black pepper, cardamom and turmeric are the major spice crops and its production and marketing are the main economy stays of lakhs of people of the country. Billions of rupees are involved in domestic trading and foreign export. Therefore, it is highly imperative for farmers to adopt the advanced practices to sustain spices productivity. Even, additional income could be generated from the intercropping of spices from the interspaces in mango orchards (Shukla and Adak, 2020). Thus the present chapter included all aspects of production to protection and value addition for the benefits of growers.

The ICAR-Indian Institute of Spices Research (ICAR-IISR), Kozhikode, Kerala, maintains the world's largest collection of black pepper, cardamom and turmeric germplasms comprising local cultivars, wild forms collected from the centres of origin and related species. The collected germplasms are characterized for various economic as well as desirable agronomic traits like yield, quality and resistance over time and the promising accessions are released as varieties. More than 70 turmeric types belonging to *Curcuma longa* are known to be under cultivation in India. Most of these cultivars are referred by local names and derived mostly from the place of occurrence (Nair *et al*., 1980). The main breeding objectives in improving these spice crops are higher yield, improvement in quality (high bulk density, dry recovery, higher levels of essential oil, oleoresin, piperine (black pepper), α-terpinyl acetate (cardamom), curcumin (turmeric), resistance to foot rot (black pepper), rhizome rot (cardamom and turmeric), resistance to nematodes (*Radopholus similis* and *Meloidogyne incognita*), resistance to insect pests especially the pollu beetle in black pepper and thrips (cardamom), resistance to drought, shade tolerance, responsive for low inputs (organic breeding), suitable for high altitude and mixed cropping (Krishnamoorthy and Parthasarathy, 2009; Ravindran and Madhusoodanan, 2002; Ravindran *et al*., 2007). The black pepper, cardamom and turmeric varieties developed by ICAR-IISR, Kozhikode for the benefits of farmers are listed in Table 1.

Table 1: Black pepper, cardamom and turmeric varieties/hybrids developed for the benefits of farmers

Breeding strategy	Varieties developed	Pedigree/ parental cultivar	Remarks
Black pepper			
Clonal selection	Sreekara	Karimunda	High quality line with high oleoresin and essential oil
	Subhakara	Karimunda	High quality line with high oleoresin and essential oil
	PLD2	Kottanadan	High oleoresin line
Selection from germplasm	Panchami	Aimpiriyan	High yielding with excellent fruit set
	Pournami	Ottaplackal	Tolerant to root knot nematode (*Meloidogyne incognita*)
	IISR Thevam	Thevanmundi	Field tolerance to Phytophthora foot rot
OP progeny selection	IISR Shakthi (P24)	Perambramundi	Tolerance to Phytophthora foot rot
	Arka Coorg Excel	-	High yielder with long spike
Inter-cultivar hybridization	IISR Malabar Excel	Cholamundi x Panniyur 1	High yield with adaptability for higher elevation
	IISR Girimunda	Narayakodi x Neelamundi	High yield with adaptability for higher elevation
Cardamom			
Clonal selection	Appangala 1	Selection from Malabar type	Suitable for intensive cultivation, produces 89% bold capsules with an average yield of 745 kg/ha

Breeding strategy	**Varieties developed**	**Pedigree/ parental cultivar**	**Remarks**
Selection from germplasm	IISR Vijetha	Germplasm selection (NKE collection)	Capsules with 77% bold capsules. Field tolerant to thrips and borer. Resistant to cardamom mosaic virus with an average yield of 643 kg/ha
OP progeny selection	IISR Avinash	OP progeny selection from CCS 1	Resistant to rhizome rot disease, produces dark green capsules with an average yield of 848 kg/ha
Inter-cultivar hybridization	Appangala 2	Hybrid (Appangala 1 x NKE 19)	Resistant to cardamom mosaic virus with an average yield 927 kg/ha
Turmeric			
Clonal selection	IISR Alleppey Supreme	Clonal selection from Alleppey turmeric	Crop duration 210 days, suitable for rainfed and irrigated condition, curcumin content (5.55%), dry recovery (19 %) with yield of 35.4 t/ha
	IISR Kedaram	Clonal selection from germplasm	Crop duration 210 days, curcumin content (5.5%), dry recovery (18. 9%) with yield of 34.5 t/ha
	IISR Pragati	Clonal selection from germplasm collections	High yield potential, short duration, curcumin content (5%) with yield of 33.19 t/ha
OP progeny selection	IISR Prabha	Open pollinated progeny selection	High yielding and good quality with reddish yellow coloured rhizome with high curcumin (6.52%) and yield of 37 t/ha
	IISR Prathibha	Open pollinated progeny selection	High quality line with curcumin (6.2%) and yield of 37.5 t/ha

Dissemination of improved technologies for the benefit of growers'

Deployment of varieties that are genetically resistant to pests/diseases provides the most efficient and economic strategy as an integral component in integrated pest/disease management. The simplest case involves complete resistance to a pathogen that remains durable for long period of time without being overcome by new virulent strains of the pathogen. The varieties of spice crops resistant against diverse biotic agents developed through different plant breeding strategies undertaken by various research centres are presented in Table 2.

Table 2: Resistant varieties in tropical spices

Black pepper			
Pest/Disease	**Variety and Institute**	**Pedigree**	**Salient features**
Phytophthora foot rot	**IISR Shakthi** ICAR - IISR, Kozhikode, Kerala	OP progeny selection from Perambramundi	Tolerant to Phytophthora foot rot. Average yield: 2253 kg dry produce/ha, Dry recovery: 43%, Piperine: 3.3%, Oleoresin: 10.2%, Essential oil: 3.7%
	IISR Thevam ICAR - IISR, Kozhikode, Kerala	Selection from Thevanmundi	Field tolerant to Phytophthora foot rot. Average yield: 5.17 kg dry produce/vine, Dry recovery: 35%, Piperine: 1.6%, Oleoresin: 8.5%, Essential oil: 3.1%
	Panniyur 8 Pepper Research Station (KAU), Panniyur, Kannur, Kerala	Hybrid of Panniyur 6 x Panniyur 5	Field tolerant to drought and Phytophthora foot rot. Average yield: 5760 kg/ha. Suited to all black pepper growing regions of Kerala
	Vijay College of Horticulture (KAU), Vellanikkara, Thrissur, Kerala	Panniyur 2 x Neelamundi	Medium long spikes and bold berries, oil, oleoresin and piperine content more than parents and Panniyur 1, field tolerant to Phytophthora foot rot
	Panniyur 9 Pepper Research Station (KAU), Panniyur, Kannur, Kerala	OP progeny selection from Panniyur 3	Spike length 10.2 cm, 90 berries per spike, 1000 berry weight 150 g, piperine: 6.11%, volatile oil: 5%, oleoresin: 12.71%. Suitable for open condition. Tolerant to foot rot and drought. Fresh yield: 7.2 kg/tree, Dry yield: 2.86 kg/tree
Root knot nematode	**Pournami** ICAR - IISR, Kozhikode, Kerala	Selection from Ottaplackal	Tolerant to root knot nematode (*Meloidogyne incognita*). Average yield: 2333 kg dry produce/ha, Dry recovery: 31%, Piperine: 4.1%, Oleoresin: 13.8%, Essential oil: 3.4%
Cardamom			
Katte/Mosaic	**IISR Vijetha** ICAR - IISR, Kozhikode, Kerala	Selection from germplasm	Resistant to *Katte* (mosaic). Yield: 643 kg dry capsules/ha, Dry recovery: 22%, Essential oil: 7.9%
	Appangala 2 ICAR - IISR, Kozhikode, Kerala	Hybrid of Appangala 1 x NKE (Natural *Katte* Escape) 19	Resistant to *Katte* (mosaic). Yield: 927 kg dry capsules/ha, Dry recovery: 21%, Essential oil: 5.8%

Black pepper			
Pest/Disease	**Variety and Institute**	**Pedigree**	**Salient features**
Rhizome rot	**IISR Avinash** ICAR - IISR, Kozhikode, Kerala	OP progeny selection of CCS 1 (Coorg Cardamom Selection 1)	Tolerant to rhizome rot. Yield: 847 kg dry capsules/ha, Dry recovery: 20.8%, Essential oil: 6.7%
	ICRI 3 ICRI (Spices Board), Myladumpara, Idukki, Kerala	Selection from Malabar type	Tolerant to rhizome rot. Yield: 440 kg under rainfed (790 kg under irrigation), Essential oil: 6.6%, Dry recovery: 22%. Early maturing, oblong, bold parrot green capsules
	ICRI 4 ICRI (Spices Board), Myladumpara, Idukki, Kerala	Clonal selection from Vadagaraparai area of lower Pulneys	Tolerant to rhizome rot. Average yield: 455 kg/ha (960 kg under irrigation)
Azhukal/ Capsule rot	**ICRI 2** Indian Cardamom Research Institute (Spices Board), Myladumpara, Idukki, Kerala	Clonal selection from germplasm collection	Tolerant to *Azhukal*/Capsule rot. Average yield: 375 kg/ha under rainfed and 766 kg under irrigated condition, Dry recovery: 22.5%.
Thrips	**Mudigere 1** Regional Research Station (UAS), Mudigere, Chikkamagaluru, Karnataka	Clonal selection from Malabar type	Moderately tolerant to thrips, hairy caterpillar and white grubs. Average yield: 275 kg dry capsules/ha, Essential oil: 8%, Dry recovery: 20%. Pale green, oval bold capsules, suitable for high density planting
	PV 2 Cardamom Research Station (KAU), Pampadumpara, Idukki, Kerala	Selection from OP seedlings of PV 1	Field tolerant to thrips. Average yield: 982 kg dry capsules/ha, Dry recovery: 23.8%, Essential oil: 10.45%. Long bold capsules
Turmeric			
Rhizome rot	**Suguna** ICAR - ICAR - IISR, Kozhikode, Kerala	Selection from germplasm	Field tolerant to rhizome rot. Average yield: 29.3t/ha, Dry recovery: 15.95%, Curcumin: 4.9%, Oleoresin: 13.5%, Essential oil: 6%, Dry recovery: 20.4%, Short duration (190 days)
	Sudarsana ICAR - ICAR - IISR, Kozhikode, Kerala	Selection from germplasm	Field tolerant to rhizome rot. Average yield: 28.8 t/ha. Early maturing

Black pepper			
Pest/Disease	**Variety and Institute**	**Pedigree**	**Salient features**
Leaf blotch	**IISR Kedaram** ICAR - IISR, Kozhikode, Kerala	Selection from germplasm	Tolerant to leaf blotch. Average yield: 34.5kg/ha, Curcumin: 5.5%, Oleoresin: 13.6%, Dry recovery: 18.9%, Duration: 210 days
	IISR Alleppey Supreme ICAR - IISR, Kozhikode, Kerala	Selection from Alleppey finger turmeric	Tolerant to leaf blotch. Average yield: 35.4 t/ha, Curcumin: 5.55%, Oleoresin: 16%, Dry recovery: 19%, Duration: 210 days
Leaf blotch, leaf spot and rhizome scale	**Suroma** High Altitude Research Station (OUAT), Pottangi, Korapurt, Odisha	Clonal selection from T. Sunder by x- ray irradiation	Field tolerant to leaf blotch, leaf spot and rhizome scale. Average yield: 20 t/ha, Curcumin: 6.1%, Oleoresin: 13.1%, Essential oil: 4.4%, Dry recovery: 26%, Duration: 253 days. Round and plumpy rhizome, suitable for Odisha, Tamil Nadu, Himachal Pradesh, Andhra Pradesh and Kerala
Leaf blotch, rhizome rot, rhizome scale and shoot borer	**Suranjana** UBKVV, Pundibari, Cooch Behar, West Bengal	Clonal selection from local types of West Bengal	Tolerant to leaf blotch and rhizome rot. Resistant to rhizome scales and moderately resistant to shoot borer. Potential yield: 29 t/ha, Curcumin: 5.7%, Oleoresin: 10.9%, Essential oil: 4.1%, Dry recovery 21.2%, Duration: 235 days. Suitable for open and shaded conditions, rainfed as well as high rainfall areas, suitable for West Bengal
Leaf blotch and Leaf spot	**Turmeric Co 3** TNAU, Coimbatore, Tamil Nadu	Selection from germplasm	Tolerant to leaf spot and leaf blotch. Curcumin: 3. 3%. Suitable for turmeric growing regions of Tamil Nadu
	Uttar Rangini UBKVV, Pundibari, Cooch Behar, West Bengal	Selection from germplasm	Tolerant to leaf spot and leaf blotch. Suitable for turmeric growing regions of West Bengal, Bihar and Tamil Nadu
Leaf spot, leaf blotch and root knot nematode	**Narendra Haldi 3** NDUAT, Kumarganj, Ayodhya, Uttar Pradesh	Selection from germplasm (NDH 9)	Resistant to root knot nematode, moderately resistant to leaf spot and leaf blotch. Average yield: 32. 5-35 t/ha. Suitable for Uttar Pradesh
Root knot nematode	**IISR Pragati** ICAR - IISR, Kozhikode, Kerala	Selection from germplasm	Moderately resistant to root knot nematodes. Average yield: 38 t/ha, Curcumin: 5.02%, Oleoresin: 15.29%, Duration: 180 days
Rhizome scale	**BSR 2** TNAU, Coimbatore, Tamil Nadu	Induced mutant from Erode local	Resistant to rhizome scale. High yielding short duration variety with bigger rhizomes

Standardizing Spices Production Technology for Growers' Benefits

Production technologies of spices crops were standardized for growers' benefits. Farmers' can grow black pepper in sandy to loam textured soils having 5.5 to 6.5 pH at an elevation up to 1500 m above sea level. It grows well at temperatures between 23°C-32°C, high humidity (75-80%) and 1250-2000 mm rainfall. For producing greater yields, management options involving high exchangeable bases, organic carbon and micronutrients are needed. Interestingly viable seeds are not generally used for propagation due lack of genetic purity; rather rooted cuttings are produced adopting several methods like traditional, rapid multiplication, trench, serpentine, vertical column and plug tray method. Among this, serpentine method is a cost effective technique to produce large number of rooted cuttings. In this method, rooted cuttings are planted in polythene bags filled with potting mixture containing top soil, sand/granite powder and FYM 2:1:1, fortified with bio-control agents (Thankamani *et al.*, 2007) which will serve as mother plants. After three months, the rooted node is further cut at each node and buried in the potting mixture to induce more roots and within 2-3 months well rooted planting materials will be ready with new sprouts. In this method the multiplication ratio is 1:60 (ICAR-IISR, 2015). In plug tray method, the rooting medium consists of soil less nursery mixture [composted coir pith and vermicompost (75:25 enriched with biocontrol agents)] and healthy rooted cuttings are ready for field planting after 120-150 days (ICAR-IISR, 2015). Vertical column method is a novel method of intensifying quality planting material. Orthotrophs can be grown on vertical column made of plastic coated welded wire mesh filled with soil less rooting mixture. In 4-5 months, about 150 single node cuttings, 10-15 laterals or plagiotrophs (to produce bush pepper) and 10 top shoots can be produced (Anandaraj, 2014). Black pepper can be planted in pits of 50 cm × 50 cm × 50 cm size dug at a spacing of 3×3 m in northern or north eastern slopes and filled with 5 kg farm yard manure, 150 g rock phosphate, 1 kg neem cake and 50 g *Trichoderma*. Rooted cuttings (2/standard) are planted 30 cm away from the base of standards like silky oak (*Grevillea*), Dadap (*Erythrina* spp.), *Garuga pinnata*, Matti (*Ailanthus malabarica*) and jack during June-July (Venkatesh *et al.*, 2020). The growing vines (1.5 m) are to be separated from standard and buried in soil to induce more leader shoots and production of lateral shoots from base of the standards. Shade regulation should be done during June and September and mulching with green leaf during the end of north east monsoon. Growing grass/legume cover like *Calopogonium mucunoides* and *Mimosa invisa* are effective soil cover to prevent soil erosion during rainy season and reduce foot rot disease incidence (Ramachandran, 1988). The critical stages of nutrient application in black pepper are flowering, fruit setting and development

(Raj, 1978). Cattle manure or compost @ 10 kg/vine and neem cake @ 1 kg/vine is to be applied before onset of south west monsoon. Application of 100:40:140 g NPK/vine is recommended in two split doses during May-June and September-October (ICAR-IISR, 2015). One-third, two-third and full dose of recommended fertilizer are to be applied during the first year, second year and third year onwards, respectively. Application of lime or dolomite @ 500 g/vine (once in two years) is recommended for acidic soil. Application of *Azospirillum* @ 50 g/vine may reduce nitrogen dose by half (AICRPS, 2005). In micronutrient deficient soil, foliar application of 0.25% zinc sulphate twice a year (May-June and September-October) and soil application of 200 g/vine magnesium sulphate is recommended (ICAR-IISR, 2015). Foliar application of micronutrient mixture (@ 5 g/L, twice) and application of ICAR-IISR Biomix consisting of PGPR (@ 100 ml/plant in nursery and 250 ml/vine) developed by ICAR-IISR is also recommended for enhanced growth and yield. Srinivasan *et al.* (2021) reported that virus infected (mild/moderately) black pepper plants could be revived by application of FYM + NPK + PGPR + micronutrients and achieved higher yield (3.6 to 7.1 kg/vine) than control (2.4 to 4.4 kg/vine). For enhancing uniform spiking and berry yield, 30 to 50 litres per vine depending on age during summer (March 15th to May 15th) at fortnightly interval irrigation is crucial. Intervention on shade regulation in February followed by irrigation (sprinkler and hose irrigation) from 1st week of March up to onset of south west monsoon at 10-15 days interval coupled with prophylactic spraying with Bordeaux mixture (1%) or carbendazim-mancozeb (0. 1%) reduced intensity of spike shedding, induced flowering and fruit setting in May-June at an altitude >1000 MSL and increased berry yield/vine. Foliar spraying of anti-transpirant like lime 1.5% showed higher photosynthetic rate and maintaining water status in leaves, thus mitigating the ill effect of climate change (Ankegowda *et al.*, 2019). Sometimes, cultivation of bush pepper in home yard is a profitable venture. Rooted lateral branches (plagiotrophs) can be grown as bush pepper either in pots or in field. Bush pepper yields throughout the year and the fresh yield can be up to 1 kg/bush after 3 years of planting (ICAR-IISR, 2015). Harvesting in black pepper started from third year onwards; December-January in plains and January-April in hilly regions. Harvest is done by hand picking the whole spikes when one or two berries turn yellow. The berries are separated and dipped in boiling water for one minute (termed as blanching to get uniform coloured black pepper) and sun dried for 3-5 days to reduce moisture content below 10%. In the process of making white pepper, fully ripened berries are immersed in water 7-8 days with frequently changing of water followed by removal of outer skin, washing, followed by drying to reach a moisture level of 12% (ICAR-IISR, 2015). Cleaning and grading operations enhance value of the produce and help to get higher returns. Cleaner machine are used to blow away the lighter fractions (dust, immature berries, pin heads and spent spikes) through a stream of air blowing system. Grading of black pepper is done using different sieves to segregate black pepper into different grades based on size. Before storage, it is to be dried to less than 10 % moisture level. The graded produce is bulk packed separately in multi-layer paper bags or woven polypropylene bags or in jute bags and stacked over the wooden pallets after laying polypropylene sheets on the floor.

For successful growing of Small cardamom, well drained, thick shady areas with loamy soil having pH 5.0 to 6.5 are preferred. The crop can be grown at an altitude of 600-1500 m above mean sea level with relative humidity of 75-90%, mean temperature of 15°C to 35°C and rainfall of 1500-2500 mm. Propagation through suckers is considered to be the most preferred method. Matured tiller with a rhizome and a growing shoot is the planting unit for establishing clonal nursery for large scale multiplication. The planting units are planted at a spacing of 1.8 m × 0.6 m (6800 plants/ha) in the trenches. Recommended fertilizers @ 48:48:96 g NPK/plant in 3 splits (starts from two months after planting), irrigation once in 15 days are recommended and shade should be avoided for better establishment (ICAR-IISR, 2015). About 30-40 suckers can be produced within one year of planting. Planting materials produced by adopting this method yields true-to-type and on the contrast, production of seedlings from seed may not be true to its parent. Small cardamom can be planted in ten months (in Karnataka) or 18 months old seedlings (in Kerala and Tamil Nadu) in June-July. Contour planting may be done in slope areas. Shade trees like Karana (*Vernonia arborea*), Corangati (*Acrocarpus fraxinifolius*), jack and Vaka (*Albizzia lebbeck*) are generally recommended (Hrideek, 2012). For planting on large scale to establish plantations, farmers should prepare pits of size 45 cm × 45 cm × 45 cm, 90 cm × 90 cm × 45 cm and 90 cm × 90 cm × 90 cm for *Malabar, Mysore* and *Vazhukka* types, respectively and filled one third with top soil, farm yard manure, 100 g rock phosphate and 50 g neem cake for better establishment of suckers. Planting diagonally along the slopes helps to reduce run off. Trench system of planting (60 cm × 30 cm) with a spacing of 2×1 m gives better establishment and greater moisture retention in soil. In sloppy areas, pits are taken along the contour. The spacing of 2. 0×2.0 m, 3.0 × 3.0 m and 2.4 × 2.4 m accommodates 2500-3000, 1111 and 1736 plants/ha in *Malabar, Mysore* and *Vazhukka* types, respectively (ICAR-IISR, 2015). To increase the yield, irrigation should be applied at 10 to 15 days interval by hose/micro-irrigation during January-May (Vijayan, 2018). In low rainfall areas, drip irrigation (4 litre/clump/day) and sprinkler irrigation (25 mm/month) may be adopted to realize better yield. Fogger/mist irrigation system should be provided in the plantations to develop a suitable microclimate for flowering and capsule setting. Growers should apply 75:75:150 kg NPK/ha under rainfed condition (2 splits) and 125:125:250 kg NPK/ha under irrigated condition (3 splits). Half of the recommended dose to be applied in soil and the remaining through foliar application is recommended to obtain significant higher yield and to save fertilizer requirements (Krishnakumar and Potty, 2002). The suitable period for soil application is during June/July and October/November and September/November/January for foliar application. One-third, two-third and full dose of recommended fertilizer are to be applied during the first year, second year and third year onwards, respectively. Application of compost @ 5 kg/plant and vermicompost @ 1 kg/plant (May/June) and agricultural lime @ 1 kg/plant (where soil pH <5. 0) is recommended (ICAR-IISR, 2015). In micronutrient deficient soil, foliar application of 0. 25% zinc sulphate (not to be mixed with any insecticide/fungicide) and soil application of boron (Borax @ 7.5 kg/ha) with NPK fertilizer is recommended in two splits during May-June and September-October. Soil pH-based foliar application of micronutrient mixture developed by ICAR-IISR specific to small cardamom is recommended (5 g/litre of water) twice, in May-June and September-October, which guarantees 15 to 25% increase in yield and quality (ICAR-IISR, 2015). Considering the surface feeding nature of small cardamom, 2 to 3 weeding at plant base is essential during May, September and December (Murugan *et al.*, 2016). Mulching should be done at 5 to10 cm thickness using fallen leaves of shade trees except during May-September to facilitate pollination by honey bees. Forking should be done at the plant base up to 90 cm and to a depth of 10 cm during November/December to promote better root penetration (ICAR-IISR, 2015). Trashing should be carried out once under rainfed condition and 2 times under irrigated condition with the onset of south west monsoon and avoid trashing after November onwards to avoid summer stress (Ashok Kumar *et al.*, 2021). Earthing up at plant base is recommended during October-December and drainage channel (45 × 30 cm) to be provided between the two rows of clumps. Cardamom gives good yield and quality under 50% shade and ensure shade lopping before onset of monsoon (May) (Alagupalamuthirsolai *et al.*, 2019). Pollination and fruit set could be enhanced by maintaining honey bee colonies @ 5/ha during the flowering season. Replanting may be taken once in 8 to10 years to maintain sustainable yield. Cardamom produces capsule 2 to 3 years after planting and harvesting started from July to December at the interval of 20 to 30 days when capsules attain physiological maturity (dark green rind and black coloured seeds). Freshly harvested capsules are washed and dried under flue curing method to get high quality green cardamom. In this method, capsules are dried up to 10 to 12% moisture level under traditional firewood. The temperature is maintained at 45 to 55°C for 3 to 4 hours under ventilation to give off the moisture from capsules, followed by maintaining the same temperature for a period of 18 to 24 hours under closed condition. In the final stage, the temperature is raised up to 60 to 65°C for another 1 to 2 hours. The dried capsules are polished either manually or using machine. The polished capsules are further graded based on the colour, density, size and percentage of immature and empty capsules. For packing, 300 gauge black polythene lined gunny bags are generally used that are subsequently stored in wooden boxes at room temperature (ICAR-IISR, 2015).

Farmers can earn good amount of profits through turmeric cultivation at their lands. Normally, land is to be prepared after the receipt of early monsoon showers with four deep ploughing and hydrated lime @ 500-1000 kg/ha to be applied for laterite soils based on soil pH level. Immediately after the receipt of pre-monsoon showers, beds of 1.0 m width, 30 cm height and of convenient length are prepared with a spacing of 50 cm between beds. Planting can also be done on ridges (ICAR-IISR, 2015). On the beds, small pits are made at a spacing of 25 cm x 30 cm and filled with well decomposed cattle manure or compost. The optimum spacing in furrows and ridges is 45-60 cm between the rows and 25 cm between the plants (ICAR-IISR, 2015). The crop is to be mulched immediately after planting with green leaves @ 12-15 t/ha. Mulching may be repeated @ 7.5 t/ha at 40 and 90 days after planting, weeding, application of fertilizers and earthing up. Weeding has to be done thrice at 60, 90 and 120 days after planting depending upon weed intensity. Farmers can grow Turmeric as an intercrop in coconut and arecanut plantations for greater benefits. It can also be raised as a mixed crop with chilli, colocasia, brinjal and cereals like maize, ragi, etc. Generally, Turmeric thrives better in well-drained sandy or clay loam soils with a pH range of 4.5-7.5 with good organic status. Turmeric can be grown in diverse tropical climates from sea level to 1500 m elevation, at a temperature range of 20°C-30°C and annual rainfall of 1500 mm or more under natural rainfall or irrigated conditions. Whole or split mother and finger rhizomes are used for planting and well developed healthy and disease-free rhizomes are to be selected. The seed rhizomes are treated with mancozeb 0.3% (3 g/L of water) for 30 minutes, shade dried for 3-4 hours and planted. A seed rate of 2500 kg rhizomes is required for planting one hectare of turmeric (Sachdeva *et al.*, 2015). A transplanting technique in turmeric using single bud sprouts (about 5 g) has been reported to enhance production of quality planting material with reduced cost. Nursery medium containing partially decomposed coir pith and vermicompost (75:25), enriched with PGPR/*Trichoderma* (10g/kg of mixture) showed conspicuous effect on turmeric bud growth. An integrated nutrient management involving a combination of inorganic fertilizers (50% N + full dose of P and K, i. e. 30-50-120 kg ha^{-1} NPK) and organic manure (20 t ha^{-1} FYM + *Bacillus megaterium* var. *Phosphaticum*) has been reported to realize higher yield level (Srinivasan *et al*., 2016). The fertilizers are to be applied in 2-3 split doses. Full dose of phosphorus is applied as basal at the time of planting. Equal split doses of N and K are top dressed at 45, 90 and 120 days after planting. In zinc deficient soils, basal application of zinc fertilizer up to 25 kg zinc sulphate/ha gives good yield. Spraying 1.4% magnesium sulphate ameliorates deficiency of magnesium. Foliar application of micronutrient mixture specific to turmeric is also recommended (dosage @ 5 g/L) twice, 60 and 90 days after planting for higher yield (ICAR-IISR, 2015). Depending upon the weather and soil conditions, about 15 to 23 irrigations are to be given in clayey soils and 40 irrigations in sandy loams (ICAR-IISR, 2015). Fertigation is to be done as per the recommended dose with 150:60:108 kg of NPK/ha and can be applied throughout the cropping period once in three days (TNAU, 2013). Growers can harvest (January-March) in seven to nine months depending on the variety and time of sowing. On maturity, the leaves turn dry and will be light brown to yellowish in colour. It is reported that, planting on June 15th recorded the highest fresh rhizome yield per hectare (47.18 t/ha) and dry rhizome yield (9. 70 t/ha). Hossain *et al.* (2005) reported direct planted turmeric yields maximum because of the larger shoot biomass production. On contrary, Shylaja *et al.* (2016) reported, planting material used as seedlings (developed through pro-tray method) recorded highest yield (67.94 kg) compared to seed rhizome in the field (43.77 kg). Rhizomes for seed purpose are generally stored by heaping in well ventilated rooms and covered with turmeric leaves. The seed rhizomes can also be stored in pits along with saw dust, sand and leaves of *Strychnos nux-vomica* (kanjiram). The rhizomes are to be dipped in quinalphos (0.075%) solution for 20-30 minutes if scale infestations are observed and in mancozeb (0.3%) to avoid storage losses due to fungal infection (Shylaja *et al.,* 2016).

Students-Scientist interaction at 2022

Training programme during 2022

Spices processing workshop at IISR

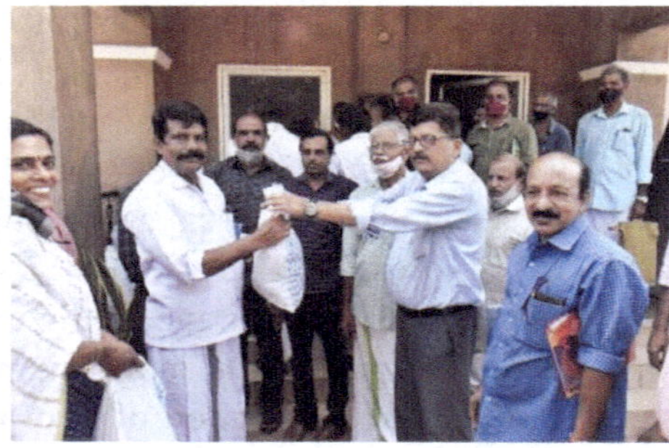

MIDH training at IISR

Standardizing Spices Protection Technology for Farmers' Benefits

Crop protection measures are essential and mandatory for obtaining sustainable spices yields. Infact, farmers have practiced a number of protection measures over decades yet several challenges to control pest menace are being faced by growers. Adoptions of modern technologies are crucial and following farm advisory at vegetative and reproductive stages are essential. Diagnosis and control management of these disease and insects are crucial for sustainable spices production. The advisory for the benefits of farmers are in table 3.

In black pepper, foot rot is caused by *Phytophthora* spp. The aerial infection occurs on the foliage, shoots, spikes and branches causing blight, spike shedding, defoliation and dieback. On the leaves, dark spots with fimbriate advancing margins or restricted lesions with yellow halo appear which spread rapidly throughout the leaves resulting in defoliation. Severe outbreaks of root rot are favoured by long duration of cloudy, humid weather with intermittent rainfall and heavy shade. Phytosanitation, providing adequate drainage, timely adoption of shade regulation and avoiding injury to the root system are vital in preventing disease initiation and subsequent spread. Application of *Trichoderma* @ 50 g/vine along with organic manures like neem cake, FYM, decomposed coffee pulp or coir pith is recommended during June-July with the onset of monsoon and during August-September. Growers are made aware about the anthracnose/Spike shedding/fungal pollu disease wich is caused by *Colletotrichum* spp. Angular to irregular brownish lesions with chlorotic halo develops on the leaves which in severe condition lead to defoliation. Brown to blackish necrotic lesions is formed on runner and orthotropic shoots and also on spikes. Regular irrigation of the vines @ 40 to 50 litres per vine at 5 to 7 days interval commencing from March and shade regulation (7500-10000 lux light) were found to reduce anthracnose incidence and spike shedding. Farmers were recommended for foliar sprays with Bordeaux mixture (1%) or carbendazim-mancozeb (0.2%) or mancozeb (0.2%) that reduces the disease initiation and subsequent spread under field conditions (Sainamole *et al.*, 2008). The symptoms of slow decline were a complex disease involving the synergistic association of nematodes and other pathogens, *Fusarium* sp. and *Phytophthora* spp. The symptoms appear as mild or moderate yellowing of the foliage. In the advanced stages, defoliation occurs; vines exhibit die-back and loss of vigour culminating in death of the vines. The vines affected with root knot nematodes exhibit dense interveinal chlorosis with galls on secondary as well as fibrous roots while primary roots get elongated swellings. Removal and destruction of the infected vines along with the root system is recommended. Growrs were also told to apply antagonistic bioagents like *Pochonia chlamydosporia* or *Trichoderma* (10^8 cfu/g) can be applied @ 50 g/vine to manage slow decline. Farmers were advocated to adopt cultural practices to recover the plants showing mild or moderate stunt symptoms with application of lime or dolomite to correct soil pH and NPK for optimum nutrient availability for the plants. Growers' should also apply plant growth promoting rhizobacteria consortia and *Trichoderma* mixed with FYM (and apply 10 to 15 kg per plant) or by drenching (2 to 3 liters per plant) two times a year (pre and post-monsoon). Spraying with micronutrient mix @ 5 g/litre twice, after spike emergence and after spike set was also suggested. Infact, stunt disease in black pepper is caused by *Piper* yellow mottle virus (PYMoV) and some of the plants infected with PYMoV show co-infection with cucumber mosaic virus (CMV). Symptoms in leaves are mosaic, yellow mottles, distortion and reduction in size. In the advanced stage, leaves get deformed, become narrow with a sickle-like appearance. The plant gets stunted due to the reduction in the intermodal length. PYMoV is transmitted through planting material and vectors (mealy bug) and CMV is mechanically transmitted. No resistant varieties are available. Use of virus-free planting materials and management of virus vectors, rouging and eradication of infected plants, removal of alternative hosts and phytosanitation are recommended. Severely infected plants should be removed and burnt or buried deep in the soil.

For controlling pest menace in small cardamom, farmers were suggested to follow a number of essentially required control measures. In case of leaf blight/ Chenthal, several species of *Colletotrichum* are reported to be associated with this disease. The disease normally appears during the mid-monsoon period (August-September) and records peak during October and November. The symptoms initially manifest on the leaves as water-soaked oblong to rectangular lesions which later elongate to form parallel arranged streaks. The lesions later turn chlorotic with a necrotic centre, which subsequently withers off. Under severe conditions, several necrotic lesions coalesce and impart a burnt-like appearance to the plants. The plant losses vigour due to excessive loss in the photosynthetic area and produce fewer flowers and panicles. If the disease appears during later stages, foliar spraying with carbendazim-mancozeb @ 0.2% or carbendazim @ 0.1% can be adopted. Association of *Pythium vexans*, *Fusarium oxysporum* and *Rhizoctonia solani* are reported with Rhizome rot disease. Initiation of the disease occurs with the onset of southwest monsoon and attains peak during July when the rainfall reaches maximum. The first visible symptoms are noticed on the leaves as pale-yellow discolouration, followed by wilting and premature necrosis of the older leaves. The characteristic symptoms are noticed near the collar regions which turn brittle and the tillers break off with mild mechanical disturbance. In advanced stages, the collar region rots, turn brownish, emits a foul smell and the affected shoots topple in the advanced stages. The emerging panicles and young shoots are also affected and further, the rotting extends to rhizomes as well as roots (Murugan *et al*., 2016). Rhizome rot tolerant varieties are IISR Avinash and ICRI 6. Removal and destruction of dead and severely affected clumps, avoiding overcrowding of plants, shade regulation and adequate drainage are recommended. Before the onset of monsoon and during August-September, drench the soil with copper oxychloride (0.25%) and foliar spray with Bordeaux mixture (1%) (Thomas and Vijayan, 1994). Drenching and spraying with potassium phosphonate (0.3%) or premixture fungicide metalaxyl-mancozeb (0.125%) both pre and post-monsoon are also recommended for growers. In order to control capsule rot in small cardamom, soil application of *Trichoderma* @1 kg during May and September to October was recommended to farmers. In addition, spraying and drenching with Bordeaux mixture (1%) and copper oxychloride (0.2%) or 2 to 3 rounds including one round of prophylactic spray with Bordeaux mixture (1%) or fosetyl aluminium (0.3%) effectively manages the disease under field conditions. Actually, the capsule rot was caused by *Phytophthora nicotianae* var. *nicotianae.* The first noticeable symptoms appear as water-soaked discoloured lesions on immature leaves and capsules which subsequently enlarge leading to decay of affected parts. The immature unopened leaves fail to unfold and as the disease spreads, the lesions turn necrotic, decay, shrivel and finally shreds. On the capsules, water-soaked discoloured patches develop which later becomes brownish leading to rotting and dropping. From the capsules, the infection may spread to the panicle thereby affecting the whole inflorescence. In advanced conditions, the entire pseudostem and panicle decay and the rotting extend to rhizomes as well as roots resulting in the collapse of the entire plant. In addition to al these diseases, virus borne disease were also significantly impacted the small cardamom production for which farmers need to follow strict prophylactic measures. Mosaic (*Katte*) disease is infected by cardamom mosaic virus

(CdMV). The infected plant is diagnosed by the presence of interveinal mosaic originating from the midrib and running towards the margin of leaves. In the advanced stage, size of the leaves gets reduced leading to stunting of the plant. Leaf sheath of infected plant exhibits mottling symptoms. From the infected plant, the disease spreads to all the tillers in the clump leading to the reduced number of small tillers with few and short panicles. The disease spreads through infected plants and from alternative hosts through the aphid, *Pentalonia caladii* in non-persistent mode. Chlorotic streak disease is incited by a strain of banana bract mosaic virus (BBrMV). The initial symptoms of the disease include spindle-shaped yellow or light green streaks intravenously. The streaks get interconnected imparting yellow or light green colour to the veins. The yellow or light green mottling is also observed on the petiole and leaf sheath of infected plants. Thereafter disease spreads to tillers in the clump severely hampering the production. The infected cardamom and banana plants serve as primary source of the virus. The secondary spread of BBrMV occurs non-persistently through aphids. The production and use of virus-free planting material is an important component in the integrated disease management package. Regular monitoring and removal of infected plants and also collateral hosts such as *Colocasia* and *Caladium* which are breeding sites of aphid are important. Vein clearing disease is caused by cardamom vein clearing virus (CdVCV). The initial symptoms of the disease consist of uninterrupted or intermittent vein clearing. Later rosetting, loosening of the leaf sheath and tearing of leaves are manifested. The newly developing leaf gets intertwined in the older leaves leading to a hook-like appearance. Infected plants are the primary source and while the secondary spread are through the cardamom aphid, *P. caladii.*

In Turmeric, rhizome rot was caused by different species of *Pythium* namely, *Pythium aphanidermatum*, *P. myriotylum* and *P. graminicolum.* Symptoms include appearance of progressive yellowing of leaves covering entire lamina leading to foliar drying. Water soaked soft lesions develop at the collar region which leads to collapse of entire plant and decay of rhizomes. Growers were advocated to grow varieties like Suvarna, Suguna and Sudarsana, and cultivar Shillong. Use of healthy rhizomes for sowing and crop rotation with maize or chilli to reduce the soil-borne inoculum, and providing adequate drainage to prevent water stagnation are important. Before sowing, treat the seeds with mancozeb (0.25%) and quinalphos (0.075%) for 15 minutes and then dry under shade. Alternatively, seed treatment followed by soil application of *Trichoderma* or PGPR multiplied in cow dung, neem cake or coir pith at the time of sowing and at regular intervals was effective for the management of rhizome rot. Farmers were also advised to drench the affected area with metalaxyl-mancozeb (0.125%) or copper oxychloride (0.25%) or mancozeb (0.2%) and repeat drenching at an interval of 15 to 20 days since the first appearance of the symptoms. Leaf blotch in turmeric appeared as pale-yellow discolouration on upper leaf surface later turning to dirty yellow and bay shade. Numerous discrete, brown to black spots coalesce and spreads to entire lamina leading to distortion and reddish-brown appearance of the leaves. Varieties like Pratibha, Rashmi, Roma, Sudharshana, Suguna, Suroma, TCP 56 and TCP 11 were recommended to use by growers. Alternatively, rhizome treatment and foliar application of carbendazim-mancozeb (0.1%) and propiconazole (0. 1%) at 45 and 90 days after planting is recommended (Sekhar and Surajit, 2014). In case of leaf spot disease, farmers were told to treat the seed rhizomes with mancozeb (0.3%) or carbendazim (0.1%) for 30 minutes and shade dry before sowing. The infected leaves should be collected and destroyed to reduce the source of inoculum in the field. When leaf spot symptoms are observed in the field, spray with mancozeb (0.25%) or copper oxychloride (0.2%) or carbendazim (0.1%) or propiconazole (0.1%) at fortnightly intervals. In general, the disease was Infection by *Colletotrichum capsici* and *C. gloeosporioides* leads to elliptic or oblong greyish white spots with yellow halo. The spots coalesce and form irregular patches which eventually dry up. Tolerant varieties like Suguna, Sudarshana Roma, Rashmi and IISR Pratibha were advocated to grow by farmers.

Table 3: Advisory services to growers for effective pest control measures in Spices

Crop	Disease	Management advisory	References
Black Pepper	Foot rot	1. Farmers were recommended to use foot rot tolerant varieties *viz.*, IISR Shakthi, IISR Thevam and Vijay in endemic regions. 2. Spraying of Bordeaux mixture (1%) and drenching the basins with copper oxychloride or metalaxyl-mancozeb (0.125% @ 5 to10 litres per vine) after commencement of monsoon are recommended.	Bhai *et al.* (2017)
Black Pepper	anthracnose/ Spike shedding/ fungal pollu	1. Growers were advocated to grow field tolerant cultivars *viz*., Aimpiriyan and Arakulammunda in endemic regions. 2. Farmers were suggested to manage the disease in nurseries with seed treatment of two/three node cuttings with carbendazim-mancozeb (0.1%) for 30 minutes and further spraying with Bordeaux mixture (1%) and carbendazim (0.1%).	Biju *et al.* (2017)

Crop	Disease	Management advisory	References
Black Pepper	Slow decline	Farmers were recommended to apply neem cake @ 2 kg per vine, addition of chopped leaves of *Glyricidia maculata* arbuscular mycorrhizal fungi, nutrient management, mulching with green manure and soil amendments like oil cakes to improve plant health.	Jasy and Koshy (1992)
Small cardamom	leaf blight / Chenthal	The disease is generally severe under less shaded conditions. Before the onset of monsoon, shade level should be maintained at 40-60% and prophylactic spray should be given with Bordeaux mixture (1%) @ 500 ml or 1 litre per plant during May-June to manage the disease.	Praveena and Biju (2012)
Small cardamom	Vein clearing virus	1. Growers were advised to use IISR Vijetha and Appangala 2 resistant to Vein clearing virus. 2. Vector load can be reduced by trashing of old and senile leaves followed by spraying with recommended insecticides or biopesticides. 3. Farmers were sensitized to use of entomopathogens like *Verticillium chlamydosporium, Beauveria bassiana* and *Paecilomyces lilacinus* would suppress aphids.	Venugopal (2002)
Turmeric	Leaf blotch	1. Use of pathogen-free seed rhizomes or treating seed rhizomes with mancozeb (0.3%) or carbendazim (0.1%) for 30 minutes followed by drying in the shade and removal and destruction of infected leaves are recommended. 2. Foliar sprays with mancozeb (0.2%) or copper oxychloride (0.25%) or propiconazole (0.1%) at fortnightly intervals are recommended to farmers.	Mishra and Pandey (2015)

References

AICRPS. (2005). Annual report. All India Coordinated Research Project on Spices. ICAR-IISR, Kozhikode, India. pp. 6-8.

Alagupalamuthirsolai, M., Ankegowda, S. J., Murugan, M., Sivaranjani, R., Balaji Rajkumar M. and Akshitha, H. J. (2019). Influence of light intensity on photosynthesis, capsule yield, essential oil and insect pest incidence of small cardamom (Elettaria cardamomum (L.) Maton). Journal of Essential Oil Bearing Plants. 22(5): 1172-1181.

Anandaraj, M. (2015). Status of plantation crops sector in India. Journal of Plantation Crops. 43(1): 1-8.

Ankegowda, S. J., Alagupalamuthirsolai, M. and Krishnamurthy, K. S. (2019). Effect of antitranspirants on gas exchange parameters and yield in black pepper during summer season. National conference of plant physiology "Plant productivity and stress management". Kerala Agricultural University, Thrissur, Kerala. Souvenir & abstracts.

Ankegowda, S J., Venugopal, M. N., Krishnamurthy, K. S. and Anandaraj, M. (2011). Impact of basin irrigation on black pepper production in coffee based cropping system in high altitude regions of Kodagu, Karnataka. Indian Journal of Horticulture. 68(1): 71-74.

Ashok Kumar, K., Murugan, M. and Dhanya, M. K. (2021). Cardamom (Elettaria cardamomum Maton). Production, breeding, management, phytochemistry and health benefits. Nova Science Publishers, Inc. New York.

Bhai, R. S., Eapen, S. J., Kumar, A., Aravind, R., Pervez, R., Varghese, E. M., Krishna, P. B. and Sreeja, K. (2017). Mitigating Phytophthora foot rot and slow decline diseases of black pepper through the deployment of bacterial antagonists. Journal of Spices and Aromatic Crops. 26: 69-82.

Biju, C. N., Praveena, R., Peeran, M. F., Darshana, C. N., Jashmi, K. C. and Ankegowda, S. J. (2017). Significance of microsclerotia in the epidemiology of black pepper anthracnose and an approach for disease management in nurseries. Journal of Phytopathology. 165: 342-353.

Hossain, M. A., Ishimine, Y., Akamine, H. and Motomura, K. (2005.) Effect of seed rhizome on growth and yield of turmeric (Curcuma longa L). Plant Production Science. 8: 86-94.

Hrideek, T. K. (2012). Shade trees and its importance in cardamom plantations. Indian Journal of Arecanut Spices and Medicinal plants. 14: 22-26.

ICAR-IISR (2015). Small cardamom-Extension Pamphlet. ICAR-Indian Institute of Spices Research, Kozhikode, Kerala, India.

ICAR-IISR. (2015). Black pepper-Extension Pamphlet. ICAR-Indian Institute of Spices Research, Kozhikode, Kerala, India.

ICAR-IISR. (2015). Turmeric-Extension Pamphlet. Published by Director, ICAR-Indian Institute of Spices Research, Calicut, Kerala, India.

Jasy, T. and Koshy, P. K. (1992). Effect of certain leaf extracts and leaves of Glyricidia maculata (H. B. and K.) Steud. as green manure on Radopholus similis. Indian Journal of Nematology. 22: 117-121.

Krishna Kumar, V. and Potty, S. N. (2002). Nutrition of cardamom. In: Ravindran, P. N. and Madhusoodanan, K. J. (Eds.) Cardamom-The genus Elettaria. Taylor and Francis, NY, pp. 129-142.

Krishnamoorthy, B. and Parthasarathy, V. A. (2009). Improvement of black pepper CAB Reviews Perspectives in Agriculture Veterinary Science Nutrition and Natural Resources. 4: 1749-8848.

Mishra, R. S. and Pandey, V. P. (2015). Management of Colletotrichum leaf spot of turmeric through fungicides. Journal of Spices and Aromatic Crops. 24: 153-156.

Murugan, M., Dhanya, M. K., Deepthy, K. B., Preethy, T. T., Aswathy, T. S., Sathyan, T., & Manoj, V. S. (2016). Compendium on Cardamom, 2nd Edition Kerala Agricultural University, Cardamom Research Station, Pampadumpara. p. 66.

Nair, M. K. Nambiar, M. C. and Ratnambal, M. J. (1980). Cytogenetics and crop improvement of ginger and turmeric. In: Nair, M. K, Premkumar, T., Ravindran, P. N. and Sarma, Y. R. (Eds.) Proceedings of National Seminar on Ginger and Turmeric, Central Plantation Crops Research Institute, Kasaragod. pp. 15-23.

Praveena, R. and Biju, C. N. (2012). Leaf blight of cardamom. Technical Bulletin, ICAR-Indian Institute of Spices Research, Kozhikode. p. 1-9.

Ramachandran, N., Sarma, Y. R., Anandaraj, M. and Abraham, J. (1988). Effect of climatic factor on Phytophthora leaf infection in black pepper grown in arecanut-black pepper mixed cropping system. Journal of Plantation Crops. 16: 110-118.

Ravindran, P. N., Babu, K. N. and Sivaraman, K. (Eds.). (2007). Turmeric: The genus Curcuma (First Edition). CRC Press. https://doi. org/10. 1201/9781420006322.

Ravindran, P. N. and Madhusoodanan, K. J. (Eds.). (2002). Cardamom: The genus Elettaria (First Edition). CRC Press. https://doi. org/10. 1201/9780203216637.

Sachdeva, N., Kumar, S. and Rana, S. S. (2015). Integrated weed management in turmeric. Indian Journal of Weed Science 47(1): 50-54.

Sainamole, P. K., Sivakumar, G., Josephrajkumar, A., Backiyarani, S., Murugan, M. and Shiva, K. N. (2008). Management of anthracnose disease (Colletotrichum glceosporioides (Penz) Penz & Sac.) of black pepper (Piper nigrum L.) in the high ranges of Idukki District, Kerala. Journal of Spices and Aromatic Crops. 17(1): 21-23.

Sekhar, B. and Surajit K. (2014). Management of foliar diseases of turmeric. Journal of Mycopathological Research. 52: 327-331.

Shukla, S. K. and Adak, T. (2020). Intercropping of turmeric for revenue generation from interspaces in mango orchards. Kahaar 7(1-2): 41-43.

Shylaja, M. R., Prasath, D. and Suresh, J. (2016). Production of quality planting material in vegetatively propagated annual spice crops-ginger and turmeric. In: Malhotra, S. K., Kandiannan, K., Mini Raj, N., Neema, V. P., Prasath, D., Srinivasan, V., Homey Cheriyan and Femina (Eds.), Proceedings of National seminar on planting material production in spices (pp. 25-40). Directorate of Arecanut and Spices Development, Kozhikode, Kerala.

Srinivasan, V., Bhat, A. I., Dinesh, R., Ankegowda, S. J., Hamza, S., Biju, C. N., Krishnamurthy, K. S., Alagupalamuthirsolai, M. and Anandaraj, M. (2021). Management of virus affected black pepper plantations through soil and plant health maintenance. Souvenir & abstracts (SYMSAC X)-2021.

Srinivasan, V., Thankamani, C. K., Dinesh, R., Kandiannan, K., Zachariah, T. J., Leela, N. K., Hamza, S., Shajina, O. and Ansha, O. (2016). Nutrient management systems in turmeric: Effects on soil quality, rhizome yield and quality. Industrial Crops and Products. 85: 241-250.

Thankamani, C. K., Dinesh, R., Eapen, S. J., Kumar, A., Kandiannan, K. and Mathew, P. A. (2007). Effect of solarized potting mixture on growth of black pepper rooted cuttings (Piper nigrum L.) in the nursery. Journal of Spices and Aromatic Crops (Supplement): 103-108.

Thomas, J. and Vijayan, A. K. (1994). Occurrence, severity, etiology and control of rhizome disease of small cardamom, Kerala Agricultural University, Vellanikkara, Thrissur.

TNAU. (2013). Turmeric. TNAU Agriportal. Tamil Nadu Agricultural University, Coimbatore, Tamil Nadu, India. https://agritech. tnau. ac. in/.

Venkatesh, L., Manju, M. J., Kavya, K. L. and Singh, K. P. (2020). Study on utility of different multipurpose trees as black pepper standards in lateritic soil of Uttara Kannada district of Karnataka, India. International Journal of Current Microbiology and Applied Science. 9(12): 2214-2223.

Venugopal, M. N. (2002). Viral diseases of cardamom. In: Ravindran, P. N. and Madhusoodanan, K. J. (Eds.). Cardamom-The genus Elettaria. Taylor and Francis. pp. 143-159.

Vijayan, A. (2018). Small cardamom production technology and future prospects. 10. 13140/ RG. 2. 2. 28724. 50567.

10

Impact of Micro-Level Agro-Advisory Services for Farming Community with Special Reference to Horticultural Crops and Extreme Events

Saon Banerjee, Abhilashaa Das, Piyali Sarkar, Sarathi Saha

Department of Agricultural Meteorology and Physics, Bidhan Chandra Krishi Viswavidyalaya, Mohanpur, Nadia, West Bengal.

Introduction

Forecasting of future weather conditions are based upon the collection and analysis of meteorological observations. With the help of advanced computing system and satellite data, the accuracy of weather forecasting is improving day by day. Weather forecasts may be classified according to the space and time scale of the predicted phenomena. With advanced technology, the spatial resolution of accurate forecast is enhanced and India Meteorological Department (IMD) has started to issue the Block-level forecast. The meteorological phenomenon can be distinguished among three classes of phenomena with a scale that is larger than the turbulent microscale: the mesoscale, synoptic scale, and planetary scale (Geer *et al.*, 2018). Considering the variability of agro-climatic condition of India, the micro-level weather based agro-advisory has great importance to enhance the farm income. IMD has initiated the Block-level Agromet Advisory Services (AAS) in collaboration with Indian Council of Agricultural Research through establishment of District Agromet Units (DAMUs) at various Krishi Vigyan Kendra's (KVK). Improvisation of the existing District-level Agromet Advisory Services (AAS) and starting of Block-level services are the basic objectives of DAMU. The District Agromet Units will act as nodal centre to prepare and disseminate the crop and location specific AAS to farmers at block level. Apart from this, the micro-level Agromet Advisory Service was implemented at NICRA (National Innovation on Climate Resilient Agriculture) villages through a network project of Indian Council of Agricultural Research (Banerjee *et al*, 2018; Chakraborty *et al.*,

2018). As a part of the nation-wide program, the feasibility of village level agromet advisory has been assessed. Thus, the importance of micro-level agromet advisory and advisory on extreme weather events are discussed in this chapter.

Micro-level Weather Based Agroadvisory

Agromet Advisory Service (AAS) deals with extension agrometeorology and is defined as 'all agrometeorological and agro-climatological information that can be directly applied to improve and/or protect the livelihood of farmers'. AAS has been adopted at district level since 2008 by the India Meteorological Department (IMD) and is continued even now (Dupdal and Naik, 2022; Vijaya Kumar *et al.*, 2017). Weather based advisory provides meteorological information such as rainfall, temperature, relative humidity and wind velocity, etc., required for farm planning and operations in shorter duration. Weather information and relevant advisories helps the farmers to take day to day farm decisions, particularly selection of crops and varieties for sowing, decision regarding intercultural operations and application of fertilizers and irrigation and decision regarding crop harvesting. Therefore, relevant weather forecast not only helps for efficient management of farm resources but also helps in enhancing crop yields and farm income. Hence, improved micro-level weather based Agromet Advisory Service (AAS) greatly helps farming community to take advantage of favourable weather and mitigate the impacts of external weather situations. In addition, weather forecast and advisories helps the farmers to increase the crop production with minimizing the losses and risks associates with farming. Timely dissemination of advisories can facilitate in reduction of farm expenditure and helps in increasing the resource use efficiency under present and climate change situations (Fig. 1). With the help of micro-level advisories, farmer can plan their farm operations well in advance and make their farming decision in shorter duration of time. The extent of risk and uncertainty associated with farming can be reduced with the help of weather advisories. However, accurate forecast and timely dissimilation of information are prerequisite for curtailing the losses arises due to weather adversities. Therefore, wide awareness and publicity among farming community for advisory service are necessary for making the full benefits out of it. The key features of this micro-level AAS include preparation and dissemination of biweekly advisories to farmers based on the forecasted weather data of IMD in consultation with KVK Subject Matter Specialists (SMS) and Field Information Facilitators (FIF) at village level. The information generated in the form of bulletin is disseminated to farmers through multi communication modes, *viz.*, mobile text and voice messages, mobile phones, WhatsApp groups, display at public places and also through personal contact. Therefore, this service can be widely disseminated and can play a crucial role in enhancing productivity through time to time advices. Time to time awareness programmes are also conducted to popularise the Agromet Advisory Services among the farming community (Fig. 2 and 3).

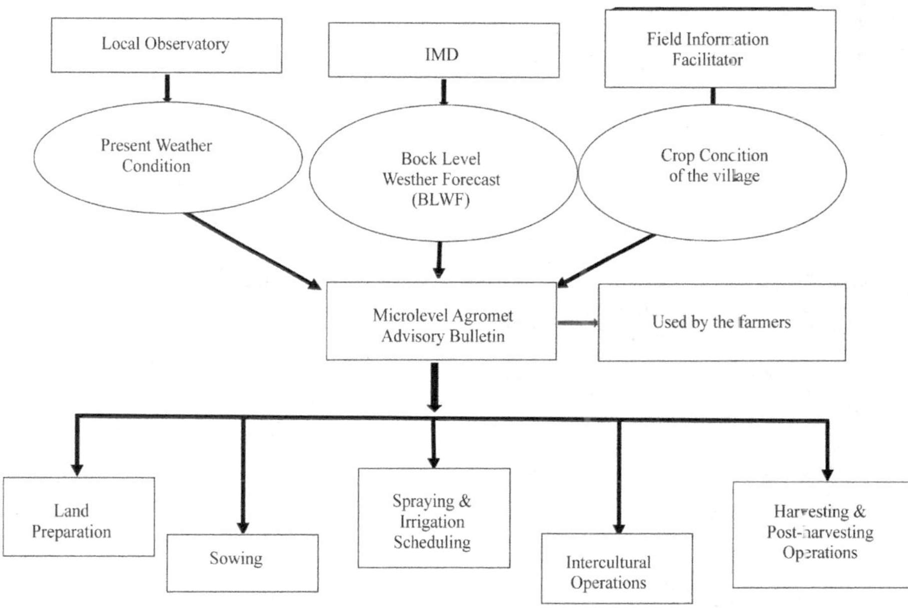

Fig. 1: Conceptual diagram of micro-level AAS and its application

Fig. 2: Farmers' Awareness Program on micro-level agroadvisory and climate change impact at Gayenpara Village, Nadia District, West Bengal

Fig. 3: Farmers awareness programme at Saluka village, Bankura District, West Bengal

Table 1: Special forecast for the cyclone

Name of the event and date of the landfall	Forecast	Expected situation
Amphan (20. 05. 2020)	Severe cyclonic storm Amphan had been intensified over the Bay of Bengal, which was classified as a "very severe cyclonic storm". When it was situated at about 900km southwest of Digha, WB, it was expected to further intensify into an "extremely severe cyclonic storm" by that night and would l touch the coastal area on 20th morning to afternoon.	• Flooding, landslides and intrusion of sea-water like Aila cyclone. • wind speed was expected to be 170-180 kmph gusting to touch even 200kmph along with heavy rainfall • Extensive damage to standing crops, plantations, orchard • Blowing down of Palm and coconut tree • Uprooting of large bushy tree
Yaas (26. 05. 2021)	The low pressure area had formed over North Andaman Sea and adjoining East Central Bay of Bengal around May 22, 2021. It was likely to intensify into a Very Severe Cyclonic Storm. According to the Met department's cyclone track, Cyclone Yaas would slowly move north-northwestwards and cross north Odisha and adjoining West Bengal coast between Paradip and Sagar around 26th. It was expected to have the wind speed ranging 155- 165 km/h during landfall. It was likely to cause heavy rainfall in the coastal districts of West Bengal and north Odisha. IMD had also warned of storm in coastal areas of West Bengal and Odisha.	• Widespread heavy rain, thunderstorm, strong wind and high waves were very likely over eastern India and adjoining areas. • Isolated extremely heavy rains were forecast over West Bengal on Tuesday and Wednesday and over Odisha on Wednesday. • Heavy to very heavy rains were likely over isolated places of Odisha on Tuesday and over isolated places of Assam, Meghalaya, Nagaland, Manipur, Mizoram, Tripura and Sikkim on Wednesday. • Heavy rain was also likely over isolated places of Assam, Meghalaya, Nagaland, Manipur, Mizoram, Tripura and Jharkhand on Tuesday, Bihar on Wednesday, and Odisha, Chhattisgarh, Jharkhand, Bihar and east Uttar Pradesh on Thursday. • Extensive damage to standing crops, plantations, orchard • Blowing down of Palm and coconut tree • Uprooting of large bushy tree • Wind speed expected to be around 155- 165 km/h during landfall

Micro-level Agroadvisory Through NICRA Project

The 'National Innovations in Climate Resilient Agriculture' (NICRA) project, launched in February, 2011, aims to enhance resilience of Indian agriculture to climate change and climate vulnerability through strategic research and technology demonstration. This network project of the Indian Council of Agricultural Research (ICAR) consists of four components *viz.* Strategic Research, Technology Demonstration, Capacity Building and Sponsored or Competitive Grants. The research on adaptation and mitigation covers crops, livestock, fisheries and natural resource management (http://www. nicra-icar. in/nicrarevised/index. php/home1). Under the umbrella of NICRA, the All India Coordinated Research Project on Agrometeorology (AICRPAM) of ICAR took up an initiative to develop and disseminate block level AAS through its 25 cooperating centres spread across the country towards enabling capacity building of farmers for climate resilience (Rao *et al.*, 2016). Considering the importance of weather and climate variability in recent years, the Central Research Institute for Dryland Agriculture (ICAR-CRIDA), Hyderabad has started a flagship research programme of the Indian Council of Agricultural Research (ICAR) named 'National Innovations in Climate Resilient Agriculture (NICRA)'. The project aims to enhance resilience of Indian agriculture to climate change and climate vulnerability through strategic research and technology demonstration. Under the broad umbrella of NICRA, the AICRPAM (All India Coordinated Research Project on Agrometeorology) took up a pilot project during 2010 to develop and disseminate block level AAS through its 25 cooperating centres throughout India. These advisories can be designated as micro-level AAS. AICRPAM centres have initiated micro-level AAS on pilot basis at more than 50 villages (minimum two villages for all the 25 centres) through the AICRPAM-NICRA project. The research fellow or scientific staff receives block level weather forecast from IMD website, and advisories are developed in consultation with Subject Matter Specialists of KVKs (25 climatic vulnerable KVKs were identified for this program). The monitory benefits from the micro-level AAS for all the villages/ centres were assessed and a significant impact was observed.

Extreme Weather Events and Need for Weather Forecasting

Extreme weather events are out of the ordinary, unexpected, unusual climatic events which wreak havoc and disrupt everyday life. Such unprecedented weather phenomena is a consequence of global climate change caused mostly by anthropogenic activities which have led to the emission of ozone depleting green-house gases like carbon dioxide, methane, nitrous oxide and chlorofluorocarbons. The frequency of occurrence of these extreme events have increased with time and laid significant impact on natural ecosystems and its habitants (Kanojia and Dijkwel, 2018). Erractic rainfall pattern (causing drought and flood), landslides, thunderstorms, heat waves and cold waves, and tropical cyclones have major impact on life and property. Agriculture, over the years, has suffered inconsiderably by such anomalous weather activities; farmers being on the receiving end. Variability in crop yield and its constant decline has caused hindrance in the economic growth of farmers (Tito *et al.*, 2018). The increased yield variability is related to increases in extreme weather events during critical growth phases of crop development. Pest and disease incidence often coincides with uncharacteristic weather conditions causing turmoil and huge loss of crops. Farming is a highly weather dependent activity. Climatic parameters have profound impact on crop growth, development and yields; on the incidence of pests and diseases; on water needs; and on fertilizer requirements. In order to adapt to or mitigate the effect of adverse weather, weather forecasting is critical. It enables farmers to cope with increasing climate uncertainty which surpasses traditional knowledge of cropping and weather patterns, making them easier to manage crops and increase productivity. Weather forecasting aids the prediction of spatial and temporal condition of atmosphere. The weather forecasting scheme operational at National Centre for Medium Range Weather Forecast (NCMRWF) for issuing location specific weather forecast five days in advance. Weather forecast and weather based agro-met advisories help in increasing the economic benefit to the farmers by suggesting them the suitable management practices according to the weather conditions (Vashisth and Singh, 2013). The relationship between weather and agriculture hasnecessitated the need for accurate prediction of the weather; to enable farmers to make an informed decision that will not bring losses to them. Temperature, sunlight, and rainfall have major effects on the crops, although, susceptibility to weather-induced stresses and affliction by pests and diseases varies amongst crops,their varieties, and different phenological stages (Didal *et al.*, 2017). The forecast of the weather event helps for suitable planning of farming operations like sowing, irrigation, fertilizer application and harvesting. Climate based strategic crop planning requires predicting of future atmospheric conditions to ensure reduction in crop failure due to extreme weather hazards. Out of all events, in the past decade, floods and cycloneshas proven to be most detrimental to crops and have caught the attention of agrarian experts. Climate induced disasters have manifolded over the years. Extreme weather events call for extreme measures. Das *et al.* (2010) in the book chapter 'Weather and Climate Forecasts for Agriculture' mentioned that "for cropping purposes, weather over short periods and year-to-year fluctuations at a particular place over the selected time interval have to be considered". Timely and accurate forecasting of weather is necessary. Medium- range weather forecasting (MRWF) is most useful for farmers in carrying out daily farming and cultural operations (Eilts, 2018).

Extreme weather events are known to destroy standing crops and cause climate-triggered pestilence. Rain-fed agriculture suffers more because of dependency towards rainfall, which gets sporadic while causing floods at times. Farmers therefore, need reliable weather forecasting to be able to be aided to climate-resilient farming and seek for crop insurance in case of inevitable failures. Climate change in India as reported by Ministry of Earth Sciences (MoES) in 2020, regions which witnessed a significant drying trend (including parts of central India, the southwest coast and north-east India), also experienced a higher annual frequency of droughts (more than two droughts per decade on average, from 1951 to 2016). Monsoon total, however, over the Indian subcontinent has been declining and is expected to decline further. Accurate and precise weather forecasting is hence a must for the hour to optimise agricultural production under changing climatic scenario. A wide range of weather forecasting methods and tools are available. Prediction of meteorological parameters like solar radiation, precipitation, bright sunshine hours, temperature, evapotranspiration, wind speed and direction, water balance, humidity and leaf wetness are important for agrometeorological applications. Apart from these, certain special agricultural weather forecasting like for frost, thunderstorms, showers, hail, tornadoes and wind gales help farmers to make timely and informed decision. However, the level of predictability of extreme events acting at meso- or microscale (frost, thunderstorms, hail, tornadoes, and so forth) must be enhanced by adopting modern technology including Artificial Intelligence (suitable for short range forecasting or nowcasting). Deep learning is being used not only for predicting usual weather patterns, it is being used to predict extreme weather conditions as well. A deep learning computer system has been able to train itself to predict accurately extreme weather conditions like heat waves or cold waves, requiring least information about current weather conditions to make predictions.

Table 2: Crop specific special advisories in view of severe cyclonic storm

Crop	Stage	Advisory given
Boro paddy	Maturity	Farmers were requested to complete harvesting and sun drying of *Boro* paddy and transfer the harvested paddy to store house immediately. It was mentioned that farm machine tools like combine harvester could be engaged for quick operation.
Ground nut, mung and black gram	Maturity	It was recommended to harvest all the matured pulses and store them in safe place after sun drying.
Summer vegetables	Fruiting	• Provision of tight bind to the cucurbits and extra support to the vegetable-scaffolding (locally called "macha") in the field was advocated so that they could withstand the high wind. • The summer vegetables were advised to be collected from the field at the earliest. • Preparation of drainage channel under the loft for removal of the excess water was also mentioned in the bulletin.
Betel vine	Green leaf	A tight bind the surroundings of Betel vine and making channel for draining out of water were proposed to the farmers.
Fruit trees	Ripening	It was stated that fruit trees like Mango, Banana, Litchi, Papaya might be severely affected, so fruit collection should be completed as soon as possible.
Cattle, chicken, duck	-	Livestock growers were opined not to let their cattle, chicken and ducklings roaming during the storm. It was also told to put them under well-built concrete shelter and arrange their food and water in sufficient amounts.

Micro-level Agroadvisory in View of Amphan and Yaas

Micro-level agroadvisory is prepared for each block of a district and disseminated among farmers twice in a week. It contains information about the upcoming weather and advisories to be taken for the standing crops and livestocks. Generally, medium range weather forecasting employing numerical weather prediction (NWP) is used to prepare the bulletins. Basically these bulletins help farmers to take necessary actions to get the probable maximum ouput as per the prevailing weather situation and also to minimize the loss due to any aberrant weather event. Along with regular one some special agromet advisories are also given at least one week prior to any extreme climatic incident based on the IMD forecasts. Recently, southern West Bengal has faced two severe cyclones, namely Amphan and Yaas. The NICRA Project played a crucial role in alerting the local people through agromet advisory dissemination. During the said severe cyclonic storms agro advisory bulletins at micro level were prepared and disseminated among the farmers through personal contact and WhatsApp group prior to at least seven days of landfall. Both of these Cyclones occurred during month of May in two consecutive years, 2020 and 2021. The Agro Advisory Bulletin (AAB) contains some general information about the upcoming events and their severity. Table 1 displays the forecast and expected situations mentioned in the bulletins for Amphan and Yaas.

The crops grown in the area were more or less at the same phenological stages in both the years. So, the advisories made for both the cyclones were similar. The crop specific advisories are given in table 2.

Some general advisories were also given as follows

- Application of chemical or irrigation to the fields in the upcoming week was strictly discouraged.
- Farmers were told that spraying application would not be a feasible operation during the cyclonic storm.
- Fishermen were advised not to venture into Bay of Bengal along & off West Bengal coasts upto the next of the prevailing cyclone.
- Farmers were requested to prepare drainage channel for removal of excess water.
- It was mentioned in the bulletin that standing under tree while lighting might put their lives at risk.
- Farmers were advised to make arrangement for food, water, urgently needed medicines, torch.
- They were given instruction to move to a safe place under adverse situation and keep important documents safely with you while moving.
- Farmers were requested to follow the Covid guidelines like maintaining social distance and wearing masks while taking all these actions.
- They were informed to stay tuned to the IMD portal and Meghdoot advisories for further details and contact with the helpline numbers if needed.

After receiving the advices the farmers took necessary actions which helped them to save the standing crops in a better way. Some advices which are to be followed by the farmers after the passage of the cyclone are also disseminated through micro-level bulletin.

Horticultural Crops and Extreme Weather Events

Unlike field crops, horticultural crops require utmost care and treatment. They are highly sensitive to weather parameters during growth stages, primarily, seedling emergence, flowering and fruit setting stage. Vulnerability to extreme weather conditions makes short duration garden crops difficult to be managed. Yield and quality of crop produce is hampered. Climate change can affect a wide range of crop production variables depending on the nature of the crop, stage of development or the time of year. Therefore, anticipation and timely response to changing climate in order to ensure both quality and quantity of horticultural crop produce is necessary. Under field circumstances, drought and heat are the most predominant stresses and have a significant influence on plants (Pereira, 2016). In tomatoes, high temperature around anthesis reduce the number of pollen grains released and their ability to germinate, both of which will have a negative impact on fruit set (Sato *et al.,* 2003). Similarly, potato crop is highly susceptible to extreme weather. There has been a growing global potato shortage owing to decline in production due to water stress, which could aggravate further. Flowers are also extremely susceptible to changes in weather indices. Extreme weather events have a greater effect on flora than presumed earlier. A month long drought can delay the time of flowering by 4 days (Jentsch *et al.,* 2008). In order to mitigate the adverse effects of extreme weather events, timely and accurate forecasting of weather along with scientific practices should be followed. Control of microclimate under protected conditions like greenhouse, polyhouse helps in quality production. Marginal and small-scale farmers may use low-cost structures like bamboo and galvanised ware. Community nursing can be followed in certain transplanted crops to combat the problem of deficit rainfall. Sowing of flowers in poly tunnel protects from frost and heavy rain in temperate regions. Emphasis should be given to the cultivation of underutilised crops like aonla, ber, carambola, spinach, wing bean etc. that are resistant to a wide range of climate and endure harsh conditions. Crop scheduling may be done based on crop type. Crops which require continuity of supply e. g. salad leaves and cauliflower, are particularly vulnerable to extreme weather events. Normally these rely on a combination of successional plantings and appropriate genotypes for reliable scheduling (Collier *et al.,* 2008). Crop weather calendars are structured based on date of sowing and variety with major emphasis on vulnerability of the crops to the changing weather conditions (Mishra *et al.,* 2011). Significant improvement can be made by genetically modifying wild species that are comparatively sturdier and less sensitive to climate change through hybridisation, selection and biotechnology. For example, frost resistant potato species like *Solanum Phureja* can be grown in case of severe frost warning. Tomato species like

Solanum chilensis, *S. peruvianum* and *S. habrochites* are disease resistant. Global climate change is often associated with pest incidence. Most pests and pathogens are temperature and humidity dependent and tend to complete more generations with higher temperatures. There are a number of models available which can be used to predict generation times, which range from relatively simple day-degree models based on linear regression to more complex process-based simulation models (Harcourt, 1954; Phelps *et al.*, 1993) and these may be used to predict the effects of climate change (Collier *et al.*, 1990). The effect of extreme climate events on the complex web of biotic interactions (pests, diseases, vectors, host plants, predators, parasitoids) is hard to assess and yet it is likely to be an important component of the outcome to crops. Pest and disease forecasting models can be used to predict the effects of extreme weather events.

Conclusion

The task of issuing time bound agroadvisory had great role in providing food security in a region. Both agricultural and horticultural crops are very much sensitive to fluctuation in meteorological parameters. Extra care, vigilance and timely action should be taken to ensure a satisfactory produce. Considering the huge variability of agro-climatic situation, the micro-level agroadvisory can enhance the overall productivity. It also helps in improving the farm income and reducing the environment pollution. India is only next to China in vegetable and fruit production. India has earned more than 550 cr from floriculture export during 2020-21. Reduction in quantity and quality of the yield of vegetables, fruits, flowers and spices by any extreme weather event is not desirable. Proper dissemination and implication of micro-level agroadvisory bulletins can play a role of saviour in this case. Scientific community along with researchers, academician, teachers, extension level officers, field staff and other trained man power are in continuous devotion to implement on ground adaptations and mitigation protocols for the benefit of farming community. This would definitely ensure the productivity and profitability for the sake of national growth.

References

A. J. Geer et al. (2018). All-sky satellite data assimilation at operational weather forecasting centres, Q. J. R. Meteorol., 144:1191–1217, DOI: https://doi. org/10. 1002/qj. 3202

Bal, S. K. and Chandran, S. (2021). Minimizing Weather-Related Risks in Agriculture Through Agromet Advisory Services in India. In Climate Change and Resilient Food Systems (pp. 245-260). Springer, Singapore.

Banerjee, S., Samanta, S. and Chakraborty, P. K. (2018). Impact of Climate Change on Costal Agro-ecosystems. Sustainable Agricultural Reviews, 33: 115-133. https://doi. org/10. 1007/978-3-319-99076-7_4

Chakraborty, A. J., Banerjee, S., Mukherjee, A., Bairagya, M. D. and Ghosh, S. (2018). Climate change and weather based Agrometeorological Advisory Service: an analysis from Farmer's Perspectives. Italian J of Agrometeorology, https://doi. org/10. 19199/2018. 3. 2038-5625. 005

Collier, R., Fellows, J. and Adams, S. (2008). Vulnerability of horticultural crop production to extreme.

Collier, R. H., Finch, S., Phelps, K. and Thompson, A. R. (1990). Possible impact of global warming on cabbage root fly (Delia radicum) activity in the UK. Annals of Applied Biology 118:261–271.

Das, H. P., Doblas-Reyes, F. J., Garcia, A., Hansen, J., Mariani, L., Nain, A. and Garcia, A. (2010). Weather and climate forecasts for agriculture. Guide to Agricultural Meteorological Practices.

Didal, V. K., Brijbhooshan, A. T. and Choudhary, K. (2017). Weather Forecasting in India: A Review. Int. J. Curr. Microbiol. App. Sci, 6(11), 577-590.

Doblas-Reyes, F., Garcia, A., Hansen, J., Mariani, L., Nain, A., Ramesh, K. and Venkataraman, R. (2003). Weather and climate forecasts for agriculture. Guide to agricultural, meteorological practices.

Dupdal, R. and Naik, B. S. (2022). Micro-Level Agromet Advisory Service: A New Approach for Minimizing Risk in Agriculture. Vigyan Varta 3(2): 99-102.

Elits, M. (2018). The Role of Weather and Weather forecasting in Agriculture

Harcourt, D. G. (1954). The biology and ecology of the diamondback moth, Plutella maculipennis (Curtis), in eastern Ontario. Ph. D. Thesis, Cornell University, Ithaca, N. Y. 107 pp.

Jentsch, A. and Beierkuhnlein, C. (2008). Research frontiers in climate change: effects of extreme meteorological events on ecosystems. Comptes Rendus Geoscience, 340(9-10), 621-628.

Kanojia, A. and Dijkwel, P. P. (2018). Abiotic stress responses are governed by reactive oxygen species and age. Annual plant reviews online, 295-326.

Mishra, M., Panda, G. R. and Padhi, B. K. (2011). Sustaining Indian agriculture in the era of climate change. In Genetics, Biofuels and Local Farming Systems (pp. 231-250). Springer, Dordrecht.

Pereira, A. (2016). Plant abiotic stress challenges from the changing environment. Frontiers in plant science, 7, 1123.

Petzoldt, C. and Seaman, A. (2006). Climate change effects on insects and pathogens. Climate change and agriculture: Promoting practical and profitable responses, 3, 6-16

Phelps, K., Collier, R. H., Reader, R. J. and Finch, S. (1993). Monte Carlo simulation method for forecasting the timing of pest insect attacks. Crop Protection, 12:335–342.

Rao, V. U. M., Bapuji Rao, B., Sarath Chandran, M. A., Vijayakumar, P. and Rao, A. V. M. S. 2016. AICRPAM–National Innovations on Climate Resilient Agriculture, Annual Report 2015–16. ICAR– Central Research Institute for Dryland Agriculture, Santoshnagar, Hyderabad, 2016, p. 52.

Sato, S. and Peet, M. M. (2005). Effects of moderately elevated temperature stress on the timing of pollen release and its germination in tomato (Lycopersicon esculentum Mill.). The Journal of Horticultural Science and Biotechnology, 80(1), 23-28.

Tito, R., Vasconcelos, H. L. and Feeley, K. J. (2018). Global climate change increases risk of crop yield losses and food insecurity in the tropical Andes. Global Change Biology, 24(2), e592-e602.

Vashisth, A., Singh, R., Das, D. K. and Baloda, R. (2013). Weather based agromet advisories for enhancing the production and income of the farmers under changing climate scenario. International Journal of Agriculture and Food Science Technology, 4(9), 847-850.

Vijaya Kumar, P., Subba Rao, A. V. M., Sarath Chandran, M. A., Venkatesh, H., Rao, V. U. M. and Srinivas Rao, Ch. (2017). Micro-level Agromet Advisory Services using block level weather forecast – A new concept-based approach. Current Science, 112(2): 227-228.

11

Extension Strategies for Technology Adoption in Horticultural Crop Growers

Naresh Babu, Tarun Adak and Arvind Kumar

ICAR- Central Institute for Subtropical Horticulture, Rehmankhera Lucknow-226101, Uttar Pradesh

Introduction

Developing countries like India are facing with the challenge to increase the agricultural productivity not only to reduce the human hunger, but also to meet the demand for nutritious food for the growing populations. Horticulture is the important sub-sector of agriculture in the country, which is playing the most important role in the livelihood of rural poor. Horticulture production in India has doubled from 146 million tonnes in 2001-02 to 320.77 million tons in 2019-20 (Verma, 2022) whereas the production of food grain has increased from 213 to 297.5 million tonnes only during the same period. In financial year 2020, fresh fruits were leading horticulture product exported from India based on value of over 54 billion Indian rupees. Keeping in view the immense potential of the horticulture in enhancing farmers' income, the Government of India allocated Rs 2,250 crore for this sector during 2021-22. In the last 2-3 decades, the horticulture sector has emerged as one of the most significant and dominant sectors not only in India but also in many other countries. Around 700,000 families are somehow involved with the horticulture sector in India. Horticulture crops including fruits are more resilient in ungracious and unfavourable weather conditions, and the vegetables exponentially improve the income of small and marginal farmers. However, government has not fixed any minimum support price for horticultural products comprising fruits such as mangoes, bananas, cucumber, pomegranate, custard apple, etc. These fruits come under the category of perishable items. As there is a lack of proper transport system and paucity of good cold chain storage, increasing the shelf-life of perishables has become a challenge. The research and development-led management along with planning strategies related to the utilization of resources in the field of horticulture would bring astounding transformation in the horticulture sector. In this direction, farm based agroadvisory to fruits and vegetable growers should be boon for farmers across region. Adoption of location specific technologies and protected cultivation were highly economical to farmers. Value-addition in horticultural products is the need of the day to meets the requirements of nutritional security. It enhances the economic value of horticultural products and unlocks new markets as well. The sector requires latest developed machinery for effective, efficient and elegant management of the horticultural crops. Many horticultural crops like aonla, pomegranate, jamun, citrus, ber, bael, field beans, cucurbits and more need insect pollinators for effective and value rich pollination to reap better outcomes. The horticulture sector has huge potential due to crop diversification factor and gives high returns compared to agricultural sector. Adak *et al.* (2018a) emphasized to follow location specific nutrient management in Dashehari mango for enhancing yield and also micronutrient practices by growers using soil test based status of orchards (Kumar *et al.,* 2015). The role of adoption of early (45 to 50 days) and high yielding Pusa Hybrid-4 bitter gourd (2.5 to 3 q/ha) for commercial cultivation in farmers field would be highly beneficial (Behera *et al.,* 2022). Likewise, Adak *et al.* (2018b) advocated adopting NPK along with 50 per cent soil and 50 percent foliar application of Zn, Cu, Mn and Fe in mango for higher yield (13.96 t/ha) and sustainable yield index (0.90). The role of extension services to disseminate technologies was thus very important and crucial and Chandra *et al.* (2019) demonstrated the positive income generation by adopting number of Farmers' Friendly Technologies. Babu *et al.* (2021a) advocated knowledge dissemination to small and marginal farmers for economic prosperity.

Fig. 1: Advisory services to farmers across region on adoption of horticultural technologies

Table 1: Farm advisory to farmers for adoption of Horticultural technologies

Horticulture system	**Extension strategies for farm advisory**	**References**
Mango	Recommended mango growers to adopt 75 per cent fertigation at critical stages for earning higher yield (over 110 kg/tree)	Adak *et al.* (2022)
Mango	Advocated use of partial root zone drying irrigation at 60 per cent crop evapotranspiration for higher yield (9 t/ha) and water productivity (6.47 kg/m^3)	Mohanty *et al.* (2022)
Cherry Tomato	Pusa Golden Cherry Tomato-2 (125 g/ha seed rate) for low cost polyhouse cultivation. Sweet and rich in nutrients, antioxidants, vitamin A and C content.	Singh *et al.* (2022)
Strawberry	Low cost polytunnel technology for successful cultivation of strawberry in semi-arid condition.	Kumar *et al.* (2016)
Pomegranate	Application of irrigation water equivalent to 70% of cumulative pan evaporation in bearing stage is necessary in Bhagwa var. pomegranate.	Marathe and Dhinesh, (2017)
Potato	Farmers were advised to grow Kufri Pukhraj and Kufri Jyoti (yield 28.70 t/ha) than use of local variety and traditional method (9.39 t/ha)	Somani *et al.* (2021)
Guava	Three times pruning in Sardar guava under 1m×2m in March, May and October to 50% shoot length for greater fruit yield (34.88 to 37.24 t/ha)	Das *et al.* (2018)
Guava	Adoption of irrigation at 75 percent of pan evaporation and 60 per cent of recommended NPK fertilizer in Lalit guava of ultra high density planting system.	Kumawat *et al.* (2017)
Okra	Application of 266, 107 and 81 kg/ha N, P_2O_5 and K_2O along with 12.5 t FYM/ha for attaining yield of 17.95 t/ha as compared to control (9.35 t/ha), only FYM (11.2t/ha) and farmers practices (13.50 t/ha).	Mageshen *et al.* (2021)
Cluster bean	Moisture conservation in 60 cm row spacing + earthen by hand plough (25 and 35 days after sowing) for higher yield (618.1 kg/ha).	Pareek *et al.* (2021)

The issuance of farm advisory from time to time helped growers for improving their orchard status. Performance of farmers' field many a times had proved beneficial by adopting newer technologies. Kumar *et al.* (2016) demonstrated at farmers field about the beneficial aspect of scientist led innovative technologies while Sureja *et al.* (2022) advocated for cultivating improved variety of round melon (Pusa Raunak) for spring-summer season.

Fig. 2: Site-specific dissemination of innovative horticultural technologies to growers

Dissemination of Resource Conservation and Management Strategies Through Demonstration and Agro-Advisory Services

The role of scientific community to dissemination of innovative technologies was always welcome for improving farmers' income. Babu *et al.* (2020) advocated for inclusion of fruit and flower cultivation in non-traditional areas for generating more income and profits through conserving natural resources. The issuance of use of liquid fertilizers like urea ammonium nitrate (UAN) for higher uptake and higher yield in vegetables may be boon for vegetable growers as urea can be saved up to 30-35 per cent (Gundappagol *et al.,* 2021). The much more aspect was to assess the soil fertility status and Vishnoi *et al.* (2020) showed 80.44 and 6.89 per cent of total geographical area of Nainital district, a hilly region in Himalayan range in the moderately fertile and less soil fertile category. Farmers accordingly follow the nutrient management practices. Likewise, soil nutrient status of other regions was also evaluated and based on which nutrient management should be adopted. The quantification of soil biological health status was also done scientifically and based on latest information, it was decided to strictly follow use of organics as biological enzymatic activities were very low (Adak and Pandey, 2020). Farmers were also advised to follow green manuring and nutrient management to improve soil and orchard sustainability (Kumar *et al.,* 2017). High density systems needs special care and nutrient cum canopy management is crucial for obtaining 15 to 16 t/ha (Adak *et al.,* 2016). In case of dry hot summers, precise water application and fertigation in fruits, vegetables and flowers were beneficial not only to mitigate adverse affect but also generate profits (Babu *et al.,* 2021b).

Cultivation of Vegetables by Adopting Mulching and Drip Irrigation

Impact of the intervention: Good numbers of farmers were cultivating few vegetables in all the FFP villages. However, they were using flood irrigation, poor and inappropriate methods in management of nutrient, pest and diseases. Hence the yield, quality and price of their produce were low. Drip irrigation system had positive improvement in water use efficiency by 90-95% over traditional flood irrigation system. There was saving of cost, labour and nutrients along with reduction in incidence of pest and diseases. All these finally led to increase in yield by 70-75% as compared to regular farmers practice. It has also resulted in increase in area under double cropping and irrigated area under vegetable crops.

Organic farming technologies: Because of increased demand for organic farm produces, farmers are to be equipped with ideas and technologies for adoption and certification of organic farming so as to get additional economic benefit of their produces. Department is encouraging organic production of Horticultural as well as field crops through MOVCD and HMNEH schemes. Pesticides used for managing pests and diseases may increase pesticide residue and, as a result, the products may be rejected or not accepted by countries. Use of recommended pesticides at the correct times and quantities may reduce pesticide residue. Farmers should employ pesticide load reducing technology (integrated pest management, washing with 2% salt water, hot water dip, bubble wash). Organically grown or pesticide residue free produce should be encouraged for health and income of farmers. The general support in this regard should be given to the adoption of "GAP" (Good Agricultural Practices). Thereby new markets can be opened for Indian mangoes. Concerted efforts are required for documentation and registration of farmers. Products shall be obtained from registered farmers to establish traceability in order to comply with importing nation standards. The lack of adequate implementation of farm-level post-harvest practices affects the shelf-life and quality of the product. For this purpose, farmers must undertake measures such as harvesting with mango harvesters at the right time and proper post-harvest management. Training and skill development in this area will improve post-harvest life of mango fruits (Ravi *et al.*, 2021).

Starts up: Location specific technologies to facilitate doubling farmers' income during 2016-17 which included enhancing farmers' income through augmentation of productivity by facilitating use of healthy seeds of improved varieties and other farm-inputs scientifically; facilitating reduction of cost of cultivation through resource use efficiency and resource conservation options; and capacity building of farmers for diversification from low income crops to high income crops and enterprises.

Crop diversification: A shift from less remunerative to more remunerative crops, more risky to less risky crops, mono to inter cropping systems have been encouraged to minimize risk in crop production, ensure reasonable returns and improve soil fertility. In mango belt areas, farmers should encourage to cultivate other fruit crops like guava, lime, bael, jamun, pineapple, custard apple, apple, apple ber, strawberry, loquat, avocado, dragon fruit, sapota, banana and vegetables and flower crops for increasing productivity per units of land and income. Turmeric, ginger and elephant foot yam in young and old orchards and could be grown for profitability. Farmers have been encouraged for crop diversification through conducting training and exposure field demonstration. In hill areas banana, mango, passion fruits, pineapple, plum, apple, peach, pear and in coastal area, pineapple, coconut, cashewnut, lemon, arecanut, banana, mango etc should be cultivated successfully.

Adoption of Protection Measures for Saving of Cost and Resources

The objective of agroadvisory services was also to save cost of cultivation and conserving natural resources. Efficient tree protection measures are useful and adopted by farmers at their fields/orchards. The beneficial role of time bound advisory was thus boon for growers across agroecological regions of the country. Weekly based advisory to farmers were helpful in managing their orchards in an effective way. Agroadvisory was issued for horticultural crops like fruits, vegetables, flowers and other crops from time to time to adopt for the benefits of farmers and over the periods around 10 lakhs farmers' were benefitted. In mango, thrips attack was serious in any agro-climatic regions. The resultant outcome is havoc fruit loss. Gundappa *et al.* (2016) scientifically analyzed the thrips population dynamics and advocated for effective management to mango growers. Farmers of vegetable (bitter gourd) growers earned over Rs. 50,000/- from half acre land with greater benefit: cost ratio of 3.1:1 and also reduced number of spray from 20 to 25 to 7 to 8 (Halder *et al.*, 2021). Strictly adoption of IPM technology in bitter gourd like seed treatment with *Trichoderma viride* @ 5g/kg of seed; spraying neem insecticide @ 5 ml/l for control of Hadda beetle; installation of cue lure traps (MAT) for fruit flies @ 10/acre; need based application of *Bacillus thuringiensis* @ 2 g/l against cucumber moth; Azadirachtin 0.03% @5 ml/l of water or NSKE 5% for whitefly; systemic fungicide against downy mildew were very efficient in reducing the incidence of pests and minimizing the yield losses. It was found that in order to control pest menace in okra, use of resistant varieties such as AE 57, PMS 8, Parkins Long Green, PKX 9275, Karnul special are highly beneficial to farmers. Growers were also advised to use *Bacillus thuringiensis* var *kurstaki* @ 500 g a. i. /ha at 10 days interval to control okra shoot and fruit borer. Alternately, use of neem extract @ 1500 ppm, if pest population is very severe was also advocated to farmers. Intercropping of soybean, maize, sorghum, cowpea with okra reduces pest infestation. Moreover, whitefly population can also be managed by using trap, botanicals and chemical insecticides of 750 ml of oxydemeton methyl 25 EC or 625 ml of dimethoate 30EC per hectare (Krishna *et al.*, 2022). Anthracnose disease in mango is dangerous for export purpose and domestic market as well. Therefore to control it, Shukla *et al.* (2017) advocated mango growers' of hot water treatment at 52±1°C for 10 minutes. Incidence of *Spodoptera* larvae were found damaging on leaves and fruits of tomato. Farmers were advised to spray Novaluron 10 EC @ 750 ml in 500 litre of water to reduce the damage of the pest or spray a biopesticide *Spodoptera* NPV @ 1ml per litre of water to control the pest. Sometimes, infestation of leaf miner *Liriomyza trifolii* was found serious on cucurbitaceous vegetables like cucumber and water melon in Malihabad area. Growers were recommended to spray 5% Neem oil instead of chemical insecticides, to control the leaf miner. In case of early sown cucurbits, red pumpkin beetle (*Raphidopalpa* sp.,) may be active and farmers were advocated dusting with 5% malathion dust at 25 kg/ha in the morning or spray malathion 50 EC @ 1.5 litre in 500 litre of water at 15 days interval to control it. Thrips (*Scirtothrips dorsalis*) and whiteflies (*Bemisia tabaci, Aleurodicus dispersus*) may cause economic loss on vegetables like chilli, tomato and brinjal. Farmers were advised to spray diafenthiuron 50 WP @ 600 g per 500-750 litre water or 5% Neem Seed Kernel Extract to effectively control of the pests. For flower growers, foliar spray of benzyl adenine @ 200 ppm twice at bud initiation stage and when flower buds were fully developed was found to be the effective treatment in improving the floral parameters in chrysanthemum (Singh and Bala, 2020).

Thus, agroadvisory services were effective in controlling the pest menace *vis-à-vis* farmers' profitability.

Fig. 3: Tree and flower protection advised to farmers for implementation and reaping benefits

Table 2: Effective horticultural advisory services for farmers' benefit

Crop	Farm advisory of extension strategies	Reference
Mango	Growing of resistant/ tolerant varieties. Early harvesting of mature fruits. As this stage of fruit maturity, crops are not susceptible to fruit fly attack. Collect and dispose all infested fruits on trees and fallen fruits on the ground every alternate day and bury in trench. Install methyl eugenol plywood traps at the rate of 6 per acre during the fruit development stage till final harvest and, if necessary, minimal application of chemical insecticides.	Megha *et. al.* (2022)
Litchi	First spray of Triazophos 40% E. C (1.5 ml/L) + Thiacloprid 21.7% SC (0.5 ml/L) or Thiacloprid 21.7% SC (0.5 ml/L) + Lambda Cyhalothrin 5% E. C (1.0 ml/L) should be done between 10th to 15th February and second spray between 25th to 28th February to control sting bug. Whenever an insecticidal spray is taken up, the fallen down insects on ground must be collected/ sweeped-off by a broom and destroyed manually by putting in a pit and covering with soils. Besides insecticidal spray, tree banding (putting sticky bands on tree trunk) may be done in the month of March-April after spraying of insecticides to prevent the nymphs to crawl up on trees. Use sticker in insecticide solution @ 0.4 ml/litre.	Kumar *et al.* (2022)
Bitter gourd	Plants affected with complex viruses should be uprooted in the early stages as well as spraying with acetamiprid 20% SP @ 100 g/ha for controlling the vectors like aphid and white fly. Fruit fly is the most harmful pest of bitter gourd which can be managed by collection and destruction of dropped fruits and infested fruits on plants, raking the soil around the plants to expose fruit fly pupae for natural enemies and placing of cue-lure traps helps in both monitoring and management of fruit fly. Spraying of spinosad @ 5 ml/10 litre of water is also effective. Collection and destruction of beetles in early stage of infestation controls red pumpkin beetle in bitter gourd.	Behera *et al.* (2022)
Winged bean	Insects such as pod borer infect pods that can be managed by spraying of cypermethrin @ 0.008 % at four different stages (flower initiation stage, 50% flowering, 100% flowering and 100% pod setting) were found effective against pod borer. Blister beetle damaged flowers can be managed by spraying of lambda-cyhalothrin @ 24g a.i. /ha twice during infestation (blooming stage).	Soni *et. al.* (2022)
Protected cultivation of tomato, brinjal, cucumber	Insect proof Nets of 40 and higher mesh are effective means of controlling most of the flying insects in brinjal, capsicum, cauliflower, tomato, okra and cucurbits. Pests and other vectors population can effectively be checked by using insect proof net-houses that creates a physical barrier between crop and open environment.	Singh *et al.* (2022)

Farmers Training

Farmers should get the recent information and technologies for adoption in their field to reap better harvest of their crops grown in commercial way. To meet up the need, department organizes farmers' training at different locations of the state and also arrange exposure visit to progressive farmers and locations outside state also.

Programmes Organized

Programme	Participants
Farmers were advised to use of pheromone trap for management of fruit fly in mango orchards. Advised to the farmers for spray of fungicide before harvesting of mango fruits to control of post harvest diseases of fruits	100
Weeding, fertilizer application and pest management in kinnow. Fungicide application in rejuvenated mango plants, advised to precise management of flowering and fruiting of mango orchards	100
Awareness among villagers regarding application of planofix for control of dropping of fruits. Irrigations should be given at 7 to 10 days intervals at pea size stage of fruits of mango. Also aware about wearing of mask during farm operation to cover nose and mouth and restrict your movement as much as possible	120
Distribution of seedlings of tomato (var. NS 285) and brinjal (var. Pusa Uttam) to farmers for developing nutri-smart gardens	80
Mechanized harvesting of mango fruits greatly improve fruit quality, reduce post harvest losses and extend the shelf life. Post harvest value addition was also advocated.	150
Adaptation of organic farming in mango orchards	100

It was observed that horticulture provides an opportunity for much needed diversification and intensification for adequate nutritious food and augmenting the income of the farmers. It is also well established that horticulture is to play a key role in doubling the farmer's income along with meeting the much-needed nutritional security. Horticulture sector should support in providing vegetables, fruits and other consumables to support the feeding of billions of people. For sustaining productivity, role of agroadvisory services are always praiseworthy and scientists and academicians are always engaged in developing innovative technologies for farmers benefit.

References

Adak, T., Kumar, K. and Singh, V. K. (2016). Variations in soil physical properties, distribution and storage of micronutrients in soil as influenced by high density mango (*Mangifera indica* L.) plantations. Journal of Soil and Water Conservation 15(1): 75-86.

Adak, T., Kumar, K. and Singh, V. K. (2018b). Sustainable yield index and soil properties in mango under different nutrient management strategy. Indian J. Hort. 75(4): 605-612.

Adak, T., Pandey, G. and Kumar, K. (2018a). Site Specific Nutrient Management in Mango - A Cursory Look. In: Soil and Water Conservation Bulletin (Mishra, P. K., Singh, L., Kumar, A., Mandal, D., Kaushal, R., Kumar, G. and Roy, T. (Eds), Indian Association of Soil and Water Conservationists, Dehradun, Uttarakhand, No. 3: 52-59.

Adak, T., Singh, V. K. and Babu, N. (2022). Optimizing the Determinants of Water Use Efficiency in Mango. Journal of Natural Resource Conservation and Management 3(1): 12-16.

Adak, T. and Pandey, G. (2020). Soil enzymatic activity as a soil health indicator for mango orchards under subtropical region. J. Soil Water Conser. 19(4): 364-369.

Babu, N., Adak, T., Arvind, K. and Singh, V. K. (2021b). Scientific analysis of drought assessment and its management strategies for food security. Current Advances in Agricultural Sciences (An International Journal) 13(1): 1-6.

Babu, N., Adak, T., Chandra, S. and Kumar, A. (2020). Crop Diversification in Fruit Belts - Sole Way to Enhance Farmers Income. In: Soil and Water Conservation Bulletin 5, Dogra, P., Sharma, N. K., Mandal, D., Kumar, G., Rawat, I., Bishnoi, R., Kar, S. K. and Kumar, S. (Eds), Indian Association of Soil and Water Conservationists, Dehradun, Uttarakhand, 2020, pp 89-94.

Babu, N., Adak, T., Pandey, G., Chandra, S. and Kumar, A. (2021a). Knowledge dissemination to small and marginal farmers for economic upliftment. In: 2nd Asian Web Conference "Managing Hill Resources and Diversities for Zero Hunger and Climate Resilience" held during 12-13 February, 2021. p 113.

Behera, T. K., Jat, G. S., Munshi, A. D. and Tomar, B. S. (2022). Pusa Hybrid-4: First gynoecious based high yielding hybrid of bitter gourd for commercial cultivation. Indian Horticulture 67(1): 3-4.

Chandra, S., Adak, T., Singh, V. K., Pandey, G., Lal, B., Babu, N. and Kumar, A. (2019). Dissemination of Farmers' Friendly Technologies Improves Farm Income. In: Soil and Water Conservation Bulletin, Mishra, P. K., Sharma, N. K., Kumar, A., Mandal, D., Kaushal, R., Kumar, G., Roy, T. and Dogra, P. (Eds), Indian Association of Soil and Water Conservationists, Dehradun, Uttarakhand, 4:73-76.

Das, B., Mehta, S., Singh, S. K., Mali, S., Dhakar, M. K. and Singh, A. K. (2018). Pruning effects on Sardar guava planted in ultra high density orcharding under different rainfall scenarios. Indian J. Hort. 75(4): 583-590.

Gundappa, Adak, T. and Shukla, P. K. (2016). Appraisal of thrips population dynamics in mango using weather based indices. Vegetos, 29(3):138-145.

Gundappagol, M., Khanna, M., Singh, M., Sudhishri, S., Rajput, J. and Dimple. (2021). Nitrogen distribution in soil under urea ammonium nitrate (UAN) and urea application in drip fertigated cabbage (*Brassica oleraceae* var. capitata). Indian Journal of Soil Conservation 49(3): 208-216.

Halder, J., Sardana, H. R., Pandey, M. K., Nagendran, K. and Bhat, M. N. (2021). Farmers' driven improved IPM module brings a better return in bitter gourd. Indian Horticulture 66(6): 10-12.

Krishna, B., Kumar, R., Choudhary, J. S., Kumar, R. and Hans, H. (2022). Insect pests in okra agro-ecosystem and their Integrated Management Indian Horticulture 67(3): 27-29.

Kumar, A., Adak, T., Muralidhara B M and Singh, V. K. (2016). Enhancing farmer's income through strawberry cultivation under subtropical climate-a new initiative. GERF Bull Biosci. 7(2):12-16.

Kumar, K., Adak, T., Lal, B., Shukla, P. K., Balaji Rajkumar, Chandra, S. and, Singh, V. K. (2016). Technological intervention in sustaining mango productivity in a resource poor farmers' field of Uttar Pradesh. GERF Bull. Biosci. 7(2):6-11.

Kumar, K., Adak, T. and Singh, V. K. (2015). Status and distribution of micronutrients in mango orchards under subtropical region of Uttar Pradesh, India. Journal of Agricultural Physics 15(2):127-139.

Kumar, K., Adak, T. and Singh, VK. (2017). Green manuring and nutrient management impacting soil properties and sustainability of mango orchard. Journal of Soil and Water Conservation 16(1): 72-78.

Kumar, V., Choudhary, J. S. and Pande, S. (2022). Litchi stink bug in Bihar: Status, threats and management strategy. Indian Horticulture 67(3): 9-11.

Kumawat, K. L., Sarolia, D. K., Kaushik, R. A. and Jodha, A. S. (2017). Effect of irrigation and fertigation scheduling on growth, flowering, yield and economics of guava cv. Lalit under ultra high density planting system. Indian J. Hort. 74(3): 362-368.

Mageshen, V. R., Ammal, U. B. and Dey, P. (2021). Influence of Soil Test Crop Response based organic manure and inorganic fertilizers on yield of okra (*Abelmoschus esculentus* L. Moench). Journal of Natural Resource Conservation and Management 2(2): 110-113.

Marathe, R. A. and Dhinesh, B. K. (2017). Plant growth, nutrient uptake, water use efficiency and yield of pomegranate as affected by irrigation scheduling in loamy soils of semi-arid regions. Indian J. Hort. 74(2): 204-209.

Megha, R., Singh, S. K., Kalia, V., Srivastav, M. and Yadav, B. (2022). Eco-friendly management of fruit fly in mango. Indian Horticulture 67(3): 27-29.

Mohanty, S., Panigrahi, P., Samant, D., Raychaudhuri, M. and Ambast, S. K. (2022). Effect of partial rootzone drying on growth and yield of mango plants. Indian Journal of Soil Conservation, 50(1): 66-71.

Pareek, N. K., Godara, S. L., Kumar, A., Rakesh, S. and Kumawat, A. (2021). Effect of moisture conservation practices on performance of cluster bean under rainfed condition. Journal of Soil and Water Conservation 20(4): 431-435.

Ravi, S. C., Rajan, S. and Verma, A. K. (2021). Status and prospects of Indian mango export industry. Indian Horticulture 66(4): 60-62.

Shukla, P. K., Adak, T. and Gundappa (2017). Anthracnose disease dynamics of mango orchards in relation to humid thermal ratio under subtropical climatic condition. J. Agrometeorol. 19(1): 56-61.

Singh, P. K., Hussian, Z., Yadav, R. K., Selvakumar, R., Mishra, I. M., Tomar, B. S., Patel, N., Yadav, S. K., Kumar, R., Kumar, J. and Kumar, A. (2022). Pusa Golden Cherry Tomato-2: New promising yellow cherry tomato for protected cultivation. Indian Horticulture 67(1): 12-15.

Singh, R. K, Rai, M., Kumar, A., Singh, A. K. and Mishra, A. C. (2022). Protected cultivation to boost the income of farmers in Bundelkhand region of Uttar Pradesh. Indian Horticulture 67(1): 36-42.

Singh, T. and Bala, M. (2020). Effect of putrescine and benzyl adenine on growth, flowering and post-harvest keeping quality parameters in chrysanthemum (*Chrysanthemum morifolium* Ramat.). J. Hortl. Sci. 15(2):191-196.

Somani, A. K., Borthakur, D., Mahant, S. and Thomas, G. (2021). Potato production with quality seed and improved technology in Assam. Indian Horticulture 66(6): 42-44.

Soni, J. K., Sunani, S. K., Lungmuana and Shakuntala, I. (2022). Winged bean: A one stop supermarket. Indian Horticulture 67(3): 35-38.

Sureja, A. K., Munshi, A. D. and Pandita, V. K. (2022). Pusa Raunak - A high yielding round melon variety for spring summer season. Indian Horticulture 67(1): 27-28.

Verma, R. (2022). Horticulture Sector: Opening up new avenues. Indian Horticulture 67(1): 2.

Vishnoi, L., Bisht, H., Shri Ram and Nain, A. S. (2020). Soil fertility assessment for ensuring food security in lower Himalayan range using GIS techniques. Indian Journal of Soil Conservation 48(1): 41-46.

12

Role of Farm Advisory for Seed Spices Growers' Benefits in India

M.D. Meena, S. Lal and S.S. Meena

ICAR-National research Center on Seed Spices, Ajmer, Rajasthan

Introduction

Farm advisories have played a key role in transforming India's agriculture, helping the country shift from a food deficit to a surplus nation. The Indian public agricultural extension system has been instrumental in this transformation through Green Revolution in the 1960s, but by the end of the 1970s, its role had started to decline. Critics argued that the system had become a monolithic organization without clear objectives (Babu *et al.*, 2013). According to NSSO data, in 2004, public extension services in India, including the Krishi Vigyan Kendra, extension workers, field demonstrations and study tours, reached to less than ten percent of farmers. Instead, farmers relied mostly on other progressive farmers (17%) and input dealers (13%) for information, while mass media including radio, television and newspaper played a significant role reaching to more than thirty percent farmers (Mkhize and Zhou, 2012; Meena *et al.*, 2022a). One of the major constraints in agricultural extension services (AES) in the country is the shortage of extension staff. According to a report, in 2012-13, a single extension functionary served 1,162 operational holdings, whereas the recommended ratio was 1:400 in hilly areas, 1:750 in irrigated areas, and 1:1000 in rain-fed areas. Input dealers often provide manipulated information specific to the crops grown by farmers, affecting their decision-making process.

To ensure the delivery of high-quality technology and information, extension systems are evolving, incorporating ICT, internet-based applications, and social platforms. These digital tools are playing a crucial role in AES globally, including in India. The use of ICT significantly impacts sustainable agricultural development and poverty reduction (Mishra *et al.*, 2020). As agricultural challenges continue to evolve, there is an increasing need for a demand-driven agricultural extension system in India (Adhiguru *et al.*, 2009). Providing farmers with timely and accurate information in a cost-effective manner is essential for agricultural development and optimal decision-making (Mittal and Mehar, 2012).

Seed Spices Cultivation Challenges

Seed spices are high-value, low-volume commercial crops extensively grown on 2.07 million hectares in India, producing 2.17 million tons during 2023-24 (Spice Board India, 2025). Rajasthan, Madhya Pradesh, and Gujarat are the major seed spice-producing states in the country contributing 53 and 50 percent to spice area and production respectively in the country during 2023-24. Seed spices hold a prominent place in India's total spice export basket and play a significant role in the national economy. Demand for seed spices has been continuously increasing. Over the past two decades, the export of seed spices has increased significantly, rising from INR 235 crore in 2000-01 to INR 7,681 crore in 2023-24 (Spice Board India, 2025), contributing 21% of the total foreign earnings from spice exports. Cumin alone contributing Rs 5,797 crore during 2023-24 with 16 percent share to total spice export from India (Spice Board India, 2025).

Seed spices are primarily grown in semi-arid and arid regions of the country, where farmers with low resource base like limited rainfall, less fertile soil with low carbon dominate. Over the last decade, climate change has emerged as a major challenge, significantly affecting overall agricultural production, particularly for smallholder farmers dependent on seed spices. Smallholder farmers are the most vulnerable to climate change due to their limited capacity to adapt. The most anticipated changes include fluctuations in rainfall and temperature, with wet areas becoming wetter and dry areas turning even drier (IPCC, 2007). Consequently, smallholder farmers in drylands, characterized by low annual rainfall (200-500 mm) and semi-arid areas (500-1000 mm annual rainfall), will experience more frequent droughts and water scarcity, further exacerbating their already water-stressed environments (El-Beltagy and Madkour, 2012; AGRA, 2014). Without the adoption of new initiatives and technologies, climate change will continue to worsen current challenges and hinder progress toward achieving the Sustainable Development Goals (SDGs) Mupanbwa *et al.,* 2020). With challenges such as climate change, population growth, deteriorating natural resources, and increasing food diversification, the role of advisory services has become even more critical (Adak *et al.*, 2018a; Babu *et al.*, 2021).

Role of Farm Advisory Services in Seed Spices

Farm advisory services act as an agricultural extension mechanism that helps farmers address their daily challenges by providing information, skills, and technology-based solutions (Anderson, 2008). Timely and well-structured farm advisory services are crucial for the economic upliftment of the rural economy, where 66% of the Indian population relies on agriculture for their livelihood. The ultimate goal of farm advisory services is to improve agricultural productivity by bridging the existing knowledge gap. These services enhance farmer incomes, promote sustainable farming practices, and contribute to long-term agricultural development. Farm advisory services help the farmers in taking informed decision on all farm related operations and activities, right from selection of variety to sale of produce in particular time, form and place. It plays a crucial role in the overall development of the agricultural sector, enhancing farm productivity, output, and earnings (Birthal *et al.*, 2015; Lal *et al.*, 2022b). These services also facilitate technology diffusion, knowledge acquisition (Feder and Slade, 1984, 1986; Babu *et al.*, 2021; Chaudhary *et al.*, 2022), and technology adoption (Cotlear, 1986; Adak *et al.*, 2018b).

Farm advisory services are delivered through multiple channels, including government agencies, agricultural extension departments, non-governmental organizations (NGOs), farmer cooperatives, private service providers, and digital platforms. Typically, these services are provided by agricultural professionals, subject matter experts, and trained extension workers with expertise in various aspects of farming. With advancements in information and communication technologies, the ways in which farm advisory services are delivered have evolved significantly over time (Torero and Braun, 2006). These services are now available through on-site visits, phone consultations, group meetings, online platforms, and mobile applications. By providing timely and customized advice, farm advisory services play a critical role in empowering farmers with sustainable and profitable farming (Fig. 1). Farm advisory services offer a wide range of support, including crop selection and planning, guidance on suitable crop choices based on agro-climatic conditions, soil type, market demand, and profitability (Fig. 2). This includes recommendations on crop rotations, intercropping, and crop diversification (Pandey *et al.*, 2018; Meena *et al.*, 2022b). Additionally, advisory services cover soil and crop management, technology adoption, market access and value addition, financial and business management, as well as training and capacity building (Adak *et al.*, 2018c; Lal *et al.*, 2022a).

Fig. 1: Farm advisory services to farmers and students training at field

Fig. 2: Awarness programme and field demonstration in seed spices.
(*Sources* for farm advisory services in seed spice sectors)

1. ICAR-National Research Centre on Seed Spice https://nrcss. icar. gov. in

The ICAR-National Research Centre on Seed Spices is an apex centre of the Indian Council of Agricultural Research working on the improvement of seed spices and the betterment of their stakeholders since its inception in 2000. This institute works for the betterment of seed spices namely coriander, cumin, fennel, fenugreek, ajwain, dill, celery, anise, nigella and caraway. The Centre is involved in various transfer of technology activities. Training for farmers, extension officers, students and other stakeholders in seed spice, National and state level farmers fairs, and National and international level seminars are organised every year. To demonstrate advanced production technologies centre organised front-line demonstrations over the country including NEH regions, on seed spices crops for better farm advisory services in the country. ICAR-NRCSS has developed Seed Portal facilities to promote the digital India programme. Farmer's corner head provides information on good agricultural practices to be followed in different seed spices and weather information for decision making. The new website has been provided with online seed purchase facilities, following self-understood step-by-step procedures. The user can track the status of the order using the link provided here. Once the order is accepted by concerned authorities, payment can also be made online using the payment gateway link provide on the website. After successful payment seeds of required crops and quantities will be delivered to the doorstep of the clients.

Source: ICAR-NRCSS website

2. Spice Board India http://www. indianspices. com

Spices Board (Ministry of Commerce and Industry, Government of India) is the flagship organization for the development and worldwide promotion of Indian spices. The Board is an international link between the Indian exporters and the importers abroad. The Board has been spearheading activities for the excellence of Indian spices, involving every segment of the industry. The board works for the Promotion of exports of spices and spice products, Maintenance and monitoring of quality of exports, Development and implementation of better production methods, through scientific, technological and economic research. Guidance to farmers on getting higher and better quality yields through scientific agricultural practices and supporting exporters for promotion of seed spice export from India.

Source: Spice Board India website

3. Mobile-based Seed Spices Info Apps

Depending on the available resources and other factors, a mobile-based scientific knowledge base app are available (Chaurasia *et al.,* 2020). It provides guidance on packages of practices in seed spices that helps farmers to cultivate these crops. Based on climate, soil type, and market need, ICT-based entrepreneurship portals can advise suitable seed spice crops for a given region. The Seed Spices Info version needs the Android 4.0.3 (IceCreamSandwich) operating system for the smartphone app SSADV, short for seed spices crop advisory, which was created using Android Studio 2.3.x. ePlatform for Seed Spices, a web-based tool, offers direction and information on many business and economic concerns connected to seed spice production, trade, and marketing. It forecasts the seed variety that provides a farmer with a higher return and, going forward, aids in determining the proper crop that should be grown by him.

4. Expert System on Seed Spices (ESSS)

A web-based expert system on seed spices called Expert System on Seed Spices (ESSS) was created. It gives all the necessary details on the nation's management of seed and spice production. The System includes 10 different seed spices in all, including cumin, fenugreek, coriander, fennel, nigella, dill and ajwain. It gives farmers of Seed Spice variety recommendations according to their locality, their cultural and climatic preferences, and other factors of relevance to the farmer. Additionally, it makes recommendations for suitable cultural practices, such as field preparation, fertiliser application, irrigation schedules, etc. It instructs them on how to safeguard the seed crop from pests, illnesses, weeds, etc. Through online enquiries, it also offers solutions to the issues experienced by farmers.

Seed Spice mobile App Expert system on seed spices
Source: ICAR-NRCSS website

5. ICAR - Indian Institute of Spices Research http://www. spices. res. in/

The Indian Institute of Spices Research (IISR), Kozhikode (Calicut) a constituent body of the Indian Council of Agricultural Research (ICAR) is a major Institute devoted to research on spices. The institute organises farmers' trainings Programmes on the Processing and evaluation of chemical quality in spices, Integrated pest and disease management in nurseries, Insect pest management, Nematode management, Biological control of diseases, Integrated nutrient management etc. The institute has developed smart spice farming solution mobile applications for users.

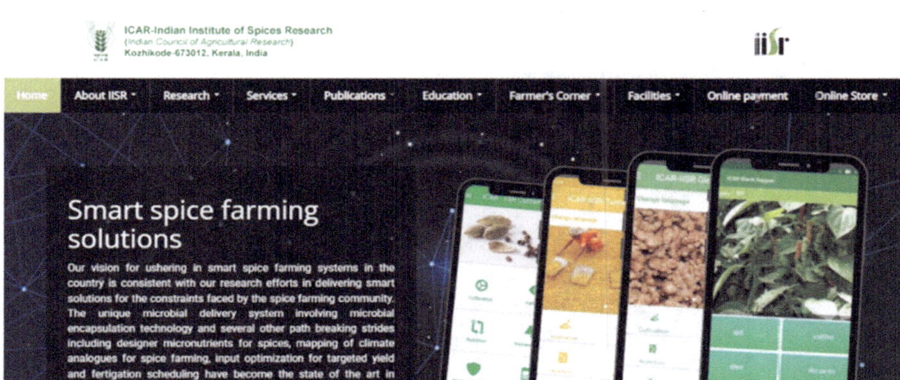

Source: ICAR-IISR website

6. Electronic National Agriculture Market (e-NAM) https://enam. gov. in/web

It is an online trading platform in India that facilitates the buying and selling of agricultural produce. eNAM was launched by the Government of India in April 2016 to create a unified national market for agricultural commodities. eNAM is designed to connect all the existing Agriculture Produce Market Committees (APMCs) or mandis in India through a common online platform. It provides a digital marketplace where farmers, traders, and buyers can interact and carry out transactions for various agricultural commodities. The platform enables farmers to discover the best prices for their products and sell directly to buyers without intermediaries, thereby eliminating the need for physical market visits and reducing transaction costs. By reducing the role of intermediaries and enabling direct farmer-buyer interactions, it seeks to empower farmers and improve their incomes. Some key features of eNAM include:

1. Online bidding and trading: eNAM allows farmers to upload information about their products, and traders can bid for the products online. This facilitates transparent price discovery and fair competition.
2. Quality certification: The platform provides quality testing and certification services for the agricultural commodities traded on eNAM. This ensures that the buyers receive goods of the specified quality.
3. Single market fee payment: Farmers can pay the market fee, commission, and other charges associated with the trading process through a single portal, irrespective of the number of markets they trade in. This simplifies the payment process and reduces paperwork.
4. Information and services: eNAM provides farmers with access to real-time market prices, historical data, demand-supply information, and weather updates. It also offers various other services like logistics, warehouse facilities, and crop insurance.

Source: https://www. enam. gov. in/web/

7. SeedNet India portal https://seednet. gov. in/

Crop productivity must be increased, which depends on the quality of the seeds. The quality of seeds is deteriorating with time, and quality control, a pricy operation, has become crucial to seed distribution and multiplication. Additionally, high-quality seeds can greatly boost farmers' and producers' bottom lines. Taking all of this into account, NIC developed the Seednet India Portal, a website that contains all information pertaining to seeds. To obtain the necessary information, farmers might visit their local distributor, government assistance centre, or NGO. Similar to this, all research institutions, certification organisations, and seed testing organisations can simply access the site for any seed-related information, including testing protocols, new crop information, etc. The portal features a variety database that offers details on different seed varieties that are now available, details on recently released seeds, etc. Its seed management database offers data on availability, supply, and demand for various seed kinds, assisting policy makers in making decisions.

8. FARMS- Farm Machinery Solutions App https://agrimachinery. nic. in/Index/farmsapp

Local farmers and residents of the various States in the country will benefit from the Custom Hiring Services of Farm Machinery Banks, Custom Hiring Centres, and Hi-tech Hubs established under the various Schemes of DAC&FW, MoA&FW with the help of this Multi-Language Mobile App Platform called "FARMS- Farm Machinery Solutions" without the need for a computer support system. This software will assist individual farmers who are prepared to rent out their agricultural machinery and equipment to boost their farm revenue while also utilising those machines to the fullest extent possible in CHCs, FMBs, and Hi-tech Machinery Hubs. This software will give farmers a place to buy and sell used agricultural equipment.

9. ICAR NRCSS Seed Portal with YONO SBI https://nrcss. icar. gov. in/SeedPortal. aspx

An innovative initiative by ICAR-NRCSS in the field of online seed sales has been taken with the integration of ICAR-NRCSS Seed Portal with YONO SBI Krishi app. Now seed growing farmers across the country will be able to buy seeds of seed crops through YONO SBI App. It should be known that State Bank YONO App is a reputed and widely used mobile banking app in India, which has access to more than five crore users of the country and is available in 10 other Indian languages besides Hindi and English. With the integration of both apps, lakhs of seed spice farmers in the country will get digital assistance in buying seeds. Farmers living in different regions of the country, who want to cultivate seed spice crops, will be able to get seeds of improved varieties of seed spices sitting comfortably in their homes. With this, the target of increasing the area and production of seed spice crops in the country will be achieved.

10. Spice Park http://www. indianspices. com/locations/spice-park

A Spice Park, also known as a Spice Processing and Packaging Cluster, is a specialized infrastructure facility that is designed to support the processing, packaging, and marketing of various spices and spice-related products. These parks are typically established in areas that are known for spice cultivation or have a significant spice industry. They aim to improve the quality of spices, enhance productivity, reduce post-harvest losses, and promote exports. Spice Parks play a crucial role in boosting the overall competitiveness of the spice industry and supporting the livelihoods of farmers and workers involved in spice cultivation and processing. Key features and facilities commonly found in Spice Parks include: processing units, warehousing and cold storage, quality control and testing laboratories, common infrastructure, research and development, and training and skill development which focus on imparting technical expertise, quality control practices, and modern processing techniques.

11. Seed Spice Atlas

An atlas is a collection of maps, charts, illustrations, and other geographical or thematic information presented as a digital resource. The primary purpose of the seed spice atlas is to provide a visual and spatial representation of India's specific seed spice cultivation areas, allowing users to gain a better understanding of the physical and geographical characteristics of different regions. With advancements in technology, digital atlases have become popular, offering interactive features, search functionalities, and the ability to zoom in and out for detailed exploration. Overall, a seed spice atlas serves as a visual guide and reference tool, enabling users to explore, analyze, and understand India's various localities of seed spice cultivation and the prospect to expand seed spice cultivation in non-traditional areas.

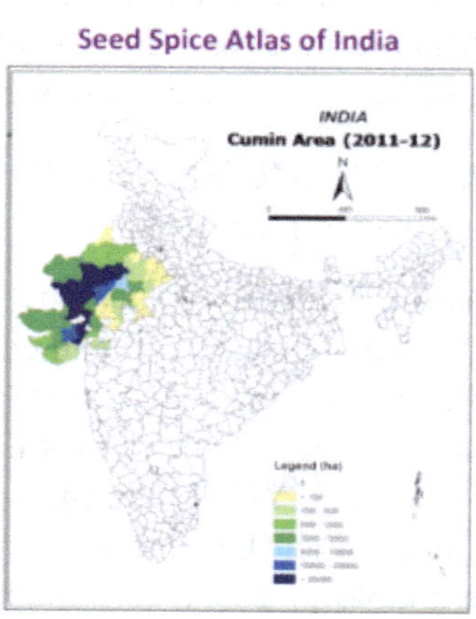

12. National Commodity and Derivatives Exchange Limited https:// www. ncdex. com/

NCDEX Limited is a Government of India online commodity and derivative exchange based in India. The national objective of exchange is the upliftment and development of farmers and the agricultural sector by connecting the primary producer to the Exchange for sustainable and inclusive growth of our agrarian economy. It is a professionally managed online for Price Discovery and Price Risk Management in major seed spices like cumin and coriander. NCDEX has 848 registered members and a client base of about 20 Lakhs and offers trading on more than 49,000 terminals across 1,000 centres in India. It facilitates deliveries of commodities through a network of over 594 accredited warehouses through eight warehouse service providers, with a holding capacity of around 1.5 million tonnes and offers average deliveries of 1 lakh MT at every contract expiry. NCDEX has offices in Mumbai, Delhi, Ahmedabad, Indore, Hyderabad, Jaipur, and Kolkata.

13. Multi Commodity Exchange of India Ltd. (MCX India)

Multi Commodity Exchange of India Ltd. is the largest commodity derivatives exchange in India. It was established in November 2003 and is headquartered in Mumbai. MCX facilitates trading in a wide range of commodities, including metals, energy products, agricultural commodities, and various other goods. As a commodity exchange, MCX provides a platform for buyers and sellers to trade in commodity futures contracts. The exchange acts as an intermediary, ensuring transparent and efficient trading, price discovery, and risk management for market participants. Trading on MCX takes place electronically through an online trading platform accessible to registered members and traders. Overall, MCX India provides a robust and transparent platform for trading in commodity futures, contributing to the development and growth of India's commodities market and providing market participants with opportunities for price discovery, hedging, and investment.

14. Participatory Guarantee System of India (PGS India)

Participatory Guarantee Systems India is a certification system for organic agricultural products in India. It is a decentralized and participatory approach to ensure the quality and integrity of organic produce. It involves the active participation of farmers, consumers, and other stakeholders in the certification process. Under PGS India, local groups or communities of farmers form a self-regulated certification system. These groups collectively define and implement the standards for organic farming practices. They conduct regular inspections, peer reviews, and documentation to ensure compliance with organic standards. The certification process in PGS India involves active involvement and verification by fellow farmers and consumers. This participatory approach builds trust and credibility within the community. PGS India certifications are recognized and accepted by the Government of India for the sale of organic products. PGS India offers an alternative to the third-party certification systems commonly used in organic agriculture. It provides an affordable and accessible certification option for small-scale farmers, especially those engaged in direct marketing and local supply chains. The PGS India certification promotes the integrity of organic products and helps consumers identify genuinely organic produce. It also empowers small farmers by reducing the cost and complexity associated with conventional certification systems. PGS India contributes to the growth and promotion of organic farming practices in India and supports the development of local and sustainable food systems.

15. e-platform

"e-platform" is the software service that delivers over the Internet the application that you need to effectively manage your enterprise. For the first time, it has been developed for agriculture-based enterprises. It is available on demand allowing a person to easily access the information available on seed spice crops regarding production technology, market information, export potential, industrial demands etc. This platform will be useful for farmers, extension workers and other stakeholders to know the developments in the area of the crop domain. The e-platform framework has been designed to speed up the development of scalable and maintainable web applications. With the advancement in technology, internet access has been provided in rural areas in all agricultural regions. The farmers can use the web for finding out about crop management, market information, export sales and historical data, as well as can know which crop he should grow in their region to obtain maximum profits (Islam *et al.*, 2014).

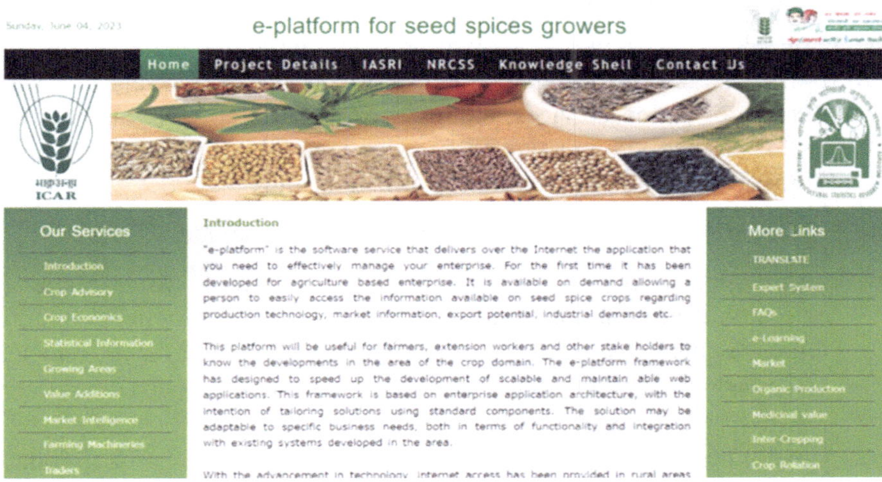

16. Buy Khari Baoli

The company, which was founded in 1962 at Asia's biggest spice market, currently takes pleasure in being the hub of wholesale commerce and handling a significant amount of trade. In this enormous market, anything is available. Both Indian and international cuisines heavily rely on spices. The foods have a fantastic texture and unparalleled flavour thanks to the spices. In this vast market, every kind of spice is available. There is a great option to purchase goods at incredibly inexpensive prices because to the market's unbelievable low rates. It strive to offer our customers high quality at a very affordable price by utilising our extensive experience. We are here to provide you with the highest quality dried fruits, premium spices, and masalas.

17. Kisan Call Centres

Kisan Call Centres, also known as KCCs, are helpline services established by the Government of India to provide agricultural information, advisory services, and assistance to farmers across the country. These call centres are specifically designed to address the needs and queries of farmers and help them with their farming-related issues. Kisan Call Centres are staffed by trained agricultural experts and professionals who possess knowledge about various aspects of farming, including crop cultivation, pest management, crop diseases, agricultural practices, market information, and government schemes and policies related to agriculture. Farmers can call the helpline number and speak directly to these experts to seek guidance and solutions for their farming problems.

18. Farmer's Portal (one stop shop for farmers)

The Farmer's Portal is a digital platform and initiative launched by the Government of India to provide a one-stop solution for farmers to access agricultural information, services, and resources. The portal serves as an online gateway for farmers, connecting them with various government schemes, agricultural programs, market-related information, and other relevant services.

The key features and functions of the Farmer's Portal include

1. Information and knowledge: The portal offers a wide range of agricultural information, including crop cultivation practices, weather forecasts, market prices, best farming practices, government policies, and schemes. Farmers can access this information to make informed decisions about their farming activities.
2. Government schemes and programs: The Farmer's Portal provides comprehensive information about various government schemes and programs available for farmers. It helps farmers understand the eligibility criteria, application processes, and benefits of these schemes, such as subsidies, crop insurance, financial assistance, and support for agricultural inputs.
3. E-marketplace: Some Farmer's Portals include e-marketplace features that allow farmers to sell their produce directly to buyers. These platforms facilitate transparent and efficient transactions, eliminating intermediaries and improving farmers' access to markets.
4. Farm management tools: The portal may offer tools and resources to help farmers manage their farms effectively. This can include crop calendars, pest and disease management guides, soil health management techniques, and other agricultural management resources.
5. Expert advice and helpline services: Some Farmer's Portals provide access to agricultural experts and helpline services. Farmers can seek advice, ask questions, and get solutions to their farming-related queries through phone, chat, or email support.
6. Access to subsidies and credit: The portal may provide information and assistance regarding agricultural subsidies, credit facilities, and loans available for farmers. It helps farmers understand the procedures and requirements for availing of financial support for agricultural activities.

19. Social Media Platforms like Facebook, WhatsApp, Twitter etc.

Social media has become a contemporary survival tool for agriculture extension in India playing a significant role in agriculture. The use of the Internet and social media has received high agenda in the extension policy at the national and international levels. Extensive use of social media like Facebook, WhatsApp, youtube, blogs etc. to transfer technologies and information in a cost-effective and timely manner can prove a fortune for agriculture extension in India. It is an instant e-extension tool, that brings more lucidity, engagement, trust and accuracy to the extension system. Social media platforms are offering numerous benefits and opportunities to farmers and the agricultural industry as a whole. These tools have become a powerful tool for farmers to access information, connect with peers, market their products, stay updated on market trends, and advocate for their interests. It enables farmers to be more informed, connected, and resilient in an increasingly digital and interconnected world.

Importance of Social Media in Agriculture

a. Information and Knowledge Sharing: Social media platforms provide a vast network for farmers to share and access valuable agricultural information, including best practices, innovative techniques, crop management strategies, and updates on new technologies. Farmers can connect with experts, researchers, and other farmers worldwide, enabling the exchange of knowledge and learning from each other's experiences.

b. Market Access and Promotion: Social media platforms offer farmers a direct and cost-effective way to promote and market their products. Farmers can showcase their produce, share product information, and engage with potential buyers, local communities, and consumer groups. This helps in building brand awareness, expanding market reach, and establishing connections with consumers interested in locally produced and sustainable agricultural products.

c. Networking and Collaboration: Social media facilitates networking and collaboration among farmers, agricultural organizations, and industry stakeholders. Farmers can join online groups, forums, and communities where they can interact, seek advice, and collaborate on various agricultural issues. This fosters the sharing of resources, ideas, and partnerships that can lead to enhanced productivity, improved practices, and collective problem-solving.

d. Access to Weather and Market Updates: Social media platforms are utilized to disseminate real-time weather forecasts, market prices, commodity trends, and other relevant information. Farmers can follow agricultural organizations, government agencies, and market analysts to stay updated on market conditions, enabling better decision-making related to crop planning, timing of sales, and pricing strategies.

e. Policy Advocacy and Awareness: Social media provides a platform for farmers to advocate for their rights, voice their concerns, and raise awareness about agricultural policies and issues. It enables farmers to connect with policymakers, agricultural organizations, and the general public to highlight the importance of agriculture, address challenges, and drive positive change in the sector.

f. Farmer Support and Empowerment: Social media creates a sense of community and support for farmers, especially those in remote or isolated areas. It offers a platform for farmers to seek assistance, share their challenges, and find solutions. It also helps in accessing resources, training opportunities, and support networks that can empower farmers to improve their livelihoods.

20. District Agro-Met Unit (DAMU)

Weather and climate information plays a vital role in agricultural production and management. Weather-advisory services are the farm decisions taken in response to past, current and future weather changes. It includes agronomical, pest and disease, water and input management. Basic considerations to prepare weather-based agro-advisories are weather-sensitive crops, their weather-sensitive stages and weather-sensitive farm operations (Fig. 3). To provide weather information and agro-advisories India Meteorological Department (IMD) in collaboration with ICAR initiated to set up District Agro-Met Unit (DAMU) at each district across India under Gramin Krishi Mausam Seva (GKMS) scheme with the following objectives:

a. To prepare and issue in time, crop-specific, weather-based District / Block level Agromet Advisory for the next 5 days by AMFUs and DAMUs
b. To collect and organize climate/weather, soil and crop information and to amalgamate them to prepare a database for value addition of advisory
c. Real-time verification of the forecast and reporting of weather events
d. To enhance the outreach and dissemination of services
e. Collection of feedback and improve the services

Fig. 3: Scientist-farmers interactions with women farmers and real time based farm advisory to farmers for seed spice production

Benefits of Farm Advisory Services to Seed Spice Stakeholders

Farm advisory services offer a range of benefits to farmers and agricultural stakeholders. Here are some key advantages of farm advisory services

1. Improved Crop Productivity: Farm advisory services provide expert guidance on crop selection, cultivation techniques, fertilization, irrigation, pest and disease management, and other agronomic practices. This helps farmers optimize their farming methods and improve crop productivity, leading to higher yields and increased profitability.
2. Enhanced Sustainability: Farm advisory services promote sustainable agricultural practices by emphasizing the use of environmentally friendly techniques and reducing reliance on chemical inputs. They provide advice on organic farming, integrated pest management, soil conservation, water management, and other sustainable practices, enabling farmers to protect natural resources and minimize negative environmental impacts.
3. Access to Modern Technologies: Farm advisory services keep farmers informed about the latest technological advancements in agriculture. They provide insights into precision farming techniques, the use of remote sensing and satellite imagery, the adoption of smart farming tools, and other innovative technologies. By integrating these technologies, farmers can optimize resource utilization, minimize losses, and improve efficiency.
4. Risk Management: Farm advisory services assist farmers in assessing and managing various risks associated with agriculture. They provide early warning systems for weather-related risks, offer insights on crop

insurance schemes, and advise on diversification strategies to mitigate market risks. This helps farmers make informed decisions and protect their crops and livelihoods from unforeseen events.

5. Market Intelligence: Farm advisory services provide market-related information, including commodity prices, demand trends, and market opportunities. They help farmers understand market dynamics, identify potential buyers or markets, and make informed decisions regarding the production, storage, and marketing of their agricultural produce. This empowers farmers to optimize their selling strategies and secure better prices for their products.
6. Capacity Building and Training: Farm advisory services offer training programs, workshops, and demonstrations to build the knowledge and skills of farmers. They guide improved agricultural practices, post-harvest management, value addition, entrepreneurship, and financial literacy. Such capacity-building initiatives enhance the technical and managerial capabilities of farmers, enabling them to adopt modern practices and manage their farms more efficiently.
7. Decision Support and Planning: Farm advisory services assist farmers in making informed decisions by providing customized advice tailored to specific farming conditions. They offer personalized farm management plans, crop calendars, and decision support tools based on soil testing, crop monitoring, and other data. This helps farmers plan their operations effectively, optimize resource allocation, and minimize uncertainties.

Overall, farm advisory services play a crucial role in supporting farmers' decision-making, enhancing productivity, promoting sustainability, managing risks, and empowering farmers with knowledge and skills to navigate the challenges of modern agriculture.

References

Adak, T., Pandey, G., Pandey, D., Garg, N., Ram Kumar, Singh, S. R., Chandra, S. and Kumar, A. (2018c). Scientist-Farmers interaction for enhancing farmers' income through nutrient management, crop diversification, value addition and other horticultural technologies. In: National Conference on Strategies & Challenges on Doubling Farmers' Income through Horticultural Technologies in Subtropics, held during June21-22, 2018 at CISH, Lucknow, pp. 115.

Adak, T., Shukla, P. K., Singh, A. K., Gundappa and Chandra, S. (2018a). A new approach to reach farmers through agroadvisory services for better orchard management. Kahaar 5(3): 31-36.

Adak, T., Singh, A. K., Chandra, S., Kumar, A. and Singh, V. K. (2018b). Improving livelihood of farmers through technology demonstration. Kahaar 5(4): 26.

Adhigurua, P., Birthal, P. S. and Kumara, G. (2009). Strengthening pluralistic agricultural information delivery systems in India. Agriculture Economics Research Review 22: 451-59.

AGRA, Alliance for a Green Revolution in Africa. (2014). Africa agriculture status report: climate change and smallholder agriculture in sub-Saharan Africa. Issue No 2. Nairobi, Kenya.

Anderson, J. R. (2008). Agricultural advisory services. Background paper for the world development report 2008. Agriculture and Rural Development Department, World Bank, Washington, DC.

Babu, N., Adak, T., Arvind, K. and Singh, V. K. (2021). Scientific analysis of drought assessment and its management strategies for food security. Current Advances in Agricultural Sciences (An International Journal) 13(1): 1-6.

Babu, N., Adak, T., Pandey, G., Chandra, S. and Kumar, A. (2021). Knowledge dissemination to small and marginal farmers for economic upliftment. In: 2nd Asian Web Conference "Managing Hill Resources and Diversities for Zero Hunger and Climate Resilience" held during 12-13 February, 2021. p 113.

Babu, S. C., Joshi, P., Glendenning, C. J., Kwadwo, A. O. and Rasheed, S. V. (2013). The state of agricultural extension reforms in India: Strategic priorities and policy options. Agricultural Economics Research Review 26(2): 159-172.

Birthal, P., Roy, D., Khan, M. T. and Negi, D. S. (2015). Farmers' preference for farming: Evidence from a nationally representative farm survey in India. The Developing Economies 53(2): 122.

Chaudhary, N., Kant, K., Dubey, P. N., Meena, S. S., Meena, R. D., Lal, S. and Saxena, S. N. (2022). Efficacy of agro Clean as pesticide in cumin and coriander. International Journal of Seed Spices 11(2): 53-57.

Chaurasia, H., Islam, S. N., Priyadarshini, S., Kumar, R. and Khan, M. A. (2020). Development of Android based Crop Advisory Application for Seed Spices. Int. J. Curr. Microbiol. App. Sci. 9(03): 904-913.

Cotlear, D. (1986). Farmer education and farm efficiency in Peru: The role of schooling, extension services and migration, Discussion Paper EDT49. Education and Training Department, Washington, DC: World Bank.

El-Beltagy, A and Madkour, M. (2012). Impact of climate change on arid lands agriculture. Agric Food Security 1:3

Feder, G. and Slade, R. (1984). The acquisition of information and the adoption of new technology. American Journal of Agricultural Economics 66(3): 312-320.

Feder, G. and Slade, R. (1986). The impact of agricultural extension: The training and visit system in India. The World Bank Research Observer 1(2): 139-161. http://documents.worldbank. org/

Islam, S. N., Bharadwaj, A, Dahiya, S., and Bhardwaj, S. P. (2014). e-Plaform for Seed Spices Growers. Project Report, ICAR-IASRI.

Lal, G., Lal, S., Chaudhary, N., Saxena, S. N., Choudhary, M. K. and Shekhawat, N. (2022b). Development of garam masala powder using organic seed spices and its storability under different packaging materials. Journal of Agriculture and Ecology 14: 77-83.

Lal, G., Lal, S., Saxena, S. N. and Choudhary, M. K. (2022a). Development and storability assessment of cumin based value added product Jeeroli. International Journal of Seed Spices 11(2): 18-22.

Meena, M., Vishal, M. K. and Meena, R. D. (2022a). Sources of information and use of mobile phones in major seed spices growing states. Guj. J. Ext. Edu. 33(2): 43-50.

Meena, S. S., Lal, S., Meena, M. D. and Lal, G. (2022b). Response of seed spices in intercropping with guava. The Pharma innovation 11(2): 690-692.

Mishra, A., Yadav, O. P., Yadav, V., Mishra, S. and Kumar, N. (2020). Benefits of the use of ICT services perceived by farmers for acquiring agricultural information in central U. P. Indian Journal of Extension Education 56 (1):86-89.

Mkhize H. P. and Zhou Y. 2012. Proceedings of The Roundtable Consultation on Agricultural Extension for Strengthening Sustainable Agriculture and Farmers' Participation in Value Chains in Asia. Available on https://www.researchgate.net/publication/266281344.

Mittal, S. and Mehar, M. (2012). How mobile phones contribute to growth of small farmers? Evidence from India. Quarterly Journal of International Agriculture 51 (3): 227-244.

Mupambwa, H. A., Ravindran, B., Dube, E., Lukashe, N. S., Katakula, A. A. N. and Mnkeni, P. N. S. (2020). Some perspectives on vermicompost utilization in organic agriculture. In: Bhat SA, Vig AP, Li F, Ravindran B (eds) Earthworm assisted remediation of effluents and wastes. Springer, Singapore, pp 299--331.

Pandey, G., Tewari, JN., Pandey, D., Shkula, SK., Trivedi, A. K., Adak, T. and Kumar, S. (2018). Possibilities of Jackfruit in the Semi arid region of India. In: National seminar on Arid Horticulture for Enhancing productivity and economic empowerment from October 27-29, 2018 at ICAR-CIAH, Bikaner, India.

Spice Board India (2021-22). http://www. indianspices. com/box2info. html

Torero, M. and Braun, J. (eds.) (2006): Information and Communication technologies for development and poverty reduction – The potential of telecommunication. The Johns Hopkins University Press and IFPRI, Washington, DC.

13

Role of Farm Advisory and Resource Management for Fruit, Vegetable and Agroforestry Growers Under Saline-Sodic Agroecosystem

Rajkumar*[1], *Mukesh Kumar Mehla*[1], *Satish Sanwal*[1], *Ashwani Kumar*[1], *Anshuman Singh*[2], *Rajkumar*[1], *Akshay*[1] *And Rakesh Banyal*[1]

[1]*Central Soil Salinity Research Institute, Karnal, Haryana*
[2]*Central Institute for Subtropical Horticulture, Lucknow, Uttar Pradesh*

Introduction

Agriculture plays a vital role in global food security and economic development. However, in many regions, agricultural productivity is hindered by various challenges, including the presence of high soil salinity in agroecological systems (Singh, 2009). Saline agroecological systems are found in various regions worldwide, including coastal areas, arid and semi-arid regions, and areas affected by improper irrigation practices. These systems are characterized by high soil salinity levels, which can have detrimental effects on crop growth, nutrient availability and plant water uptake (Rai *et al.*, 2020). Addressing these challenges requires specialized knowledge and tailored strategies to ensure sustainable agricultural practices in such environments (Dagar *et al.*, 2019; Sethi *et al.*, 2016). Farm advisory and resource management services play a critical role in assisting farmers in saline agroecological systems by providing expert guidance, information dissemination, capacity building, and policy support (Adak *et al.*, 2020). Farm advisory can play a crucial role in helping farmers understand the intricacies of saline agroecological systems, including the causes of soil salinity, its effects on crop performance, and the necessary strategies for managing these challenges effectively. Farm advisory can assist farmers in selecting suitable crop varieties that exhibit tolerance to high salinity levels. This requires comprehensive knowledge of salt-tolerant crops and their specific requirements, including their adaptability to saline conditions and nutritional needs. Providing information on the performance and market demand of different crop varieties enables farmers to make informed decisions about crop selection, optimizing productivity in saline agroecological systems (Kumar *et al.*, 2020; Singh *et al.*, 2021). It can offer guidance on crop management practices tailored to saline environments by providing recommendations on irrigation scheduling, water management techniques, and appropriate fertilization strategies to minimize salt accumulation and maximize crop yields. Helping farmers implement best practices for saline agroecological systems, ensuring optimal crop growth and productivity.

Water and soil management are critical aspects of agriculture in saline agroecological systems. The proper advisory can support farmers in adopting efficient irrigation practices that minimize salt buildup and optimize water use efficiency. By application of irrigation methods such as drip irrigation, which delivers water directly to the root zone, reducing salt leaching and water wastage (Mehla *et al.*, 2022) is important to develop irrigation schedules under micro irrigation. In this study, effect of different spacing between drip lateral (45 cm and 60 cm). In terms of soil management, more emphasis should be given to the importance of soil amendments and organic matter incorporation to improve soil structure and fertility. We also promote the application of gypsum or other soil amendments to reduce soil salinity levels. The use of organic materials and cover crops to enhance soil health, increase organic matter content, and improve water-holding capacity in saline agroecological systems. Advisory associated with pest and disease management can provide farmers with information on pests and diseases that are prevalent in these environments and offer strategies for integrated pest management (Gundappa *et al.*, 2016a; Shukla *et al.*, 2017). Institute have been advocating for cultural practices, biological control methods, and the judicious use of pesticides to minimize the impact of pests and diseases on crops. Farm advisory services not only focus on production aspects but also extend support to market access and value chain development. Facilitating market linkages and promoting value chain development, helping farmers' access profitable markets, thereby enhancing their economic sustainability (Singh *et al.*, 2018; Babu *et al.*, 2022). This chapter aims to explore the role of farm advisory and resource management in agriculture within saline agroecological systems, highlighting the importance of understanding the unique characteristics of these environments and adopting effective management practices.

Challenges Under Saline-Sodic Agroecological Systems

Saline and sodic agroecosystems are prevalent in various regions of India and soil area in different states of India is presented in Fig. 1. Natural processes, such as seawater intrusion, saltwater irrigation, and inadequate drainage, contribute to the formation and degradation of these ecosystems. Additionally, human activities, including improper irrigation practices, excessive use of chemical fertilizers, and inadequate soil and water management, further exacerbate the challenges faced in saline and sodic agroecosystems. The presence of salts in irrigation water and high evaporation rates contribute to salt accumulation in the soil. Excessive salt buildup hampers water uptake, impairs nutrient absorption, and reduces photosynthetic efficiency. Additionally, limited water availability and poor water quality further exacerbate the challenges faced by farmers in these ecosystems. Plants cultivated in saline and sodic agroecosystems encounter various physiological and metabolic challenges. The high salt concentration disrupts the osmotic balance within plant cells, causing water stress and reduced growth. Salt stress affects nutrient uptake, leading to nutrient imbalances and deficiencies. Moreover, salt accumulation alters the soil microbial community, impacting nutrient cycling and plant-microbe interactions. As a result, farmers in these regions face numerous difficulties in sustaining profitable agricultural practices. Salinity-induced yield losses are observed in a wide range of crops, fruits, vegetables, and cash crops (Haj-Amor *et al.*, 2022; Tuomisto *et al.*, 2017). Crop sensitivity to salinity varies, with some plants exhibiting higher tolerance and others being more susceptible. The economic implications of reduced crop productivity are severe, as farmers face diminished incomes and increased vulnerability. Understanding the specific challenges encountered in saline and sodic agroecosystems is essential for developing effective strategies and management practices to mitigate their adverse effects on agriculture.

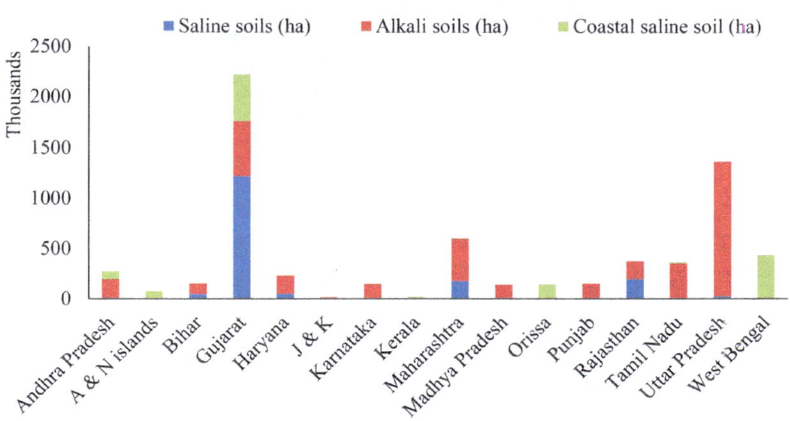

Fig. 1: Salinity-sodicity affected areas in different states of India (https://cssri. res. in/)

Fig. 2: Scientist-farmers interaction and training to farmers for Horti-Agroforesty production in problematic soils

Fruit Cultivation Under Saline-Sodic Agroecological Systems

Fruit production is an essential component of the agricultural sector, contributing significantly to the country's economy and food security. However, fruit growers in many regions face significant challenges; especially cultivating fruits in saline-sodic ecosystems is a big challenge. Salinity and sodicity can severely hamper fruit production and quality (Gora *et al.*, 2017). Therefore, understanding the dynamics of fruit production in saline-sodic ecosystems is crucial for developing effective strategies to enhance productivity and ensure sustainable fruit farming practices. High soil salinity affects various physiological processes of fruit plants, including water uptake, nutrient absorption, and photosynthesis. As a result, fruit trees experience water stress, nutrient deficiencies, reduced growth rates, and increased susceptibility to diseases and pests. These factors collectively lead to decreased fruit yields and compromised fruit quality, with implications for both growers and consumers (Barman *et al.*, 2021b; Rajkumar *et al.*, 2020; Singh *et al.*, 2023). Different fruit crops exhibit varying degrees of tolerance to saline-sodic stress. Some fruit crops, such as date palm (*Phoenix dactylifera*) and pomegranate (*Punica granatum*), have a relatively higher tolerance to salinity and can be grown in moderately saline soils (Banyal *et al.*, 2019). On the other hand, fruit crops like citrus (*Citrus* spp.) and grapes (*Vitis vinifera*) are more sensitive to high soil salinity and require careful management practices to mitigate the negative impacts. Understanding the crop-specific responses to saline-sodic stress is vital for developing targeted management strategies and selecting appropriate fruit varieties for cultivation in these challenging environments (Fig. 2).

Vegetable Cultivation Under Saline-Sodic Agroecological Systems

Vegetable production plays a crucial role in the agricultural sector, providing a significant source of nutrition and income for millions of people. Saline-sodic conditions significantly impact vegetable production. Understanding the dynamics of vegetable production in saline-sodic ecosystems is crucial for developing effective strategies to enhance productivity and ensure sustainable vegetable farming practices. High soil salinity interferes with essential physiological processes of vegetable plants, such as water absorption, nutrient uptake, and photosynthesis. Consequently, vegetable crops experience water stress, nutrient imbalances, reduced growth rates, and increased vulnerability to diseases and pests. These factors collectively lead to decreased vegetable yields, compromised nutritional quality, and economic losses for farmers (Sanwal *et al.*, 2021, 2019). Different vegetable crops exhibit varying degrees of tolerance to saline-sodic stress. Some vegetable crops, such as spinach (*Spinacia oleracea*) and okra (*Abelmoschus esculentus*), have a relatively higher tolerance to salinity and can be grown in moderately saline soils. On the other hand, vegetable crops like tomato (*Solanum lycopersicum*) and cucumber (*Cucumis sativus*) are more sensitive to high soil salinity and require careful management practices to mitigate the negative impacts (Banyal *et al.*, 2019; Tuomisto *et al.*, 2017). Integrated approaches encompassing soil, water, and nutrient management and pest and disease control strategies are crucial for optimizing vegetable production under saline-sodic conditions.

Agroforestry Under Saline-Sodic Agroecological Systems

Agroforestry, the integrated management of trees, crops, and livestock on the same piece of land, holds significant potential for sustainable agriculture. However, agroforestry systems face unique challenges when implemented in saline-sodic ecosystems, where high levels of salts and sodium negatively impact soil fertility, water availability, and plant growth. The high salinity levels affect tree establishment, growth, and productivity, leading to reduced yields, stunted growth, and increased susceptibility to pests and diseases. Additionally, the excess exchangeable sodium causes soil compaction, restricts root development, and affects nutrient uptake by both trees and associated crops. Agroforestry systems offer several potential benefits in saline-sodic ecosystems. The presence of trees can mitigate the adverse effects of salinity by lowering the water table, reducing soil waterlogging, and enhancing soil organic matter content. Agroforestry can also improve soil structure, increase nutrient cycling, carbon sequestration and provide shade and windbreak, creating a more favourable microclimate for plant growth (Adak *et al.*, 2018). Furthermore, the combination of trees, crops, and livestock in agroforestry systems can enhance productivity, diversify income sources, and contribute to environmental sustainability. Choosing tree species that are tolerant to salinity and sodicity is crucial for successful agroforestry. Species such as *Casuarina equisetifolia* and *Prosopis juliflora* have shown promising tolerance to saline conditions in India (Jinger *et al.*, 2023; Sharma and Singh, 2019). Farm advisory for the selection of appropriate tree species based on their adaptability to saline-sodic soils and their compatibility with associated crops is essential for ensuring the sustainability and productivity of agroforestry systems. Intercropping with salt-tolerant crops, such as millets or legumes, can provide additional income while utilizing resources efficiently (Dagar, 2013; Gupta and Dagar, 2016). Advocating incorporation of nitrogen-fixing legumes which can improve soil fertility by enhancing nitrogen availability, selection of salt-tolerant tree species, soil improvement practices, efficient water management, and appropriate crop selection are critical considerations in achieving successful agroforestry in saline-sodic ecosystems.

Role of Farm Advisory and Recommendations

Farm advisory is crucial in supporting farmers and ensuring sustainable agricultural practices. These services provide expert guidance, information, and resources to help farmers make informed decisions and overcome various challenges. In saline agroecological systems, where high soil salinity poses significant challenges to agricultural productivity, the role of farm advisory services becomes critical. Farm advisors play a pivotal role in assisting farmers in understanding the unique characteristics of saline environments and implementing effective strategies to mitigate the negative impacts of soil salinity. Farm advisory services encompass a range of activities aimed at assisting farmers in making informed decisions about their farming operations. These services are typically provided by agricultural extension officers, agronomists, and other experts who possess specialized knowledge in different aspects of farming. Farm advisors offer guidance on crop selection, soil management, irrigation practices, pest and disease control, market access, and other vital areas relevant to agricultural production (Gundappa *et al.*, 2016b). The role of farm advisory and resource management in agriculture is invaluable for the success and sustainability of farming operations. Through expert guidance, knowledge dissemination, capacity building, and policy support, farm advisors empower farmers to make informed decisions, adopt best practices, and overcome challenges. Farm advisory services cover a wide range of areas critical for agricultural productivity and sustainability, from crop selection and soil management to water conservation, pest control, market access, and climate change adaptation (Adak *et al.*, 2021). Different aspects of farm advisory are depicted in Fig. 3.

Weather Information Systems

The role of weather information systems in farm advisory services has become increasingly vital in modern agriculture. With the advent of advanced technology and meteorological forecasting, farmers can now access real-time weather data and predictions that enable them to make informed decisions regarding crop management practices. In addition, weather information systems provide valuable insights into temperature, precipitation, humidity, wind patterns, and other meteorological factors that directly impact crop growth, pest and disease incidence, and overall farm productivity. Traditionally, farm advisory services relied on conventional methods and local knowledge to guide farmers on various agricultural practices. However, with technological advancements, weather information systems have emerged as indispensable tools for farm advisory services. These systems utilize weather data collected from meteorological stations, satellites, and other sources, combined with advanced modelling techniques, to generate forecasts and insights specific to agricultural needs.

Crop Selection and Diversification

Farm advisory services assist farmers in selecting appropriate crop varieties based on factors such as soil type, climate conditions, market demand, and farmer preferences. Advisors provide information on the performance, suitability, and potential risks associated with different crops, helping farmers make informed decisions that maximize their chances of success. Crop diversification in the traditional area with monoculture is now becoming reality for enhancing farmers' income (Babu *et al.,* 2020). Additionally, farm advisors offer recommendations on crop management practices, including quality planting materials, planting techniques, nutrient management, crop rotation, and harvesting methods, to optimize yields and ensure sustainable farming practices.

Soil Management and Conservation

Effective soil management is critical for maintaining soil health and productivity. Farm advisors provide guidance on soil testing, nutrient management, organic matter application, and erosion control techniques. In addition, they help farmers implement sustainable soil management practices, such as conservation tillage, cover cropping, and precision agriculture, to minimize soil degradation, enhance nutrient availability, and promote long-term soil health management (Adak and Pandey, 2021).

Pest and Disease Management

Pests and diseases can significantly impact farm productivity and quality. Farm advisory services help farmers implement integrated pest and disease management strategies that minimize chemical inputs while maximizing control effectiveness. Advisors emphasize the use of cultural practices, biological control methods, and resistant crop varieties to reduce reliance on pesticides and promote environmentally friendly pest management approaches. By providing real time information on pest and disease outbreaks, monitoring techniques, and appropriate control measures, farm advisors enable farmers to make informed decisions and protect their crops effectively (Barman *et al.*, 2021a; Gundappa *et al.,* 2016c).

Post-Harvest Handling and Market Access

Post-harvest activities, including processing, storage, and handling, are critical stages that directly impact agricultural produce's quality, market value, and shelf life. Farm advisory services play a crucial role in guiding farmers on best practices and technologies to ensure optimal post-harvest management.

Post-harvest processing and handling involve a range of activities, including cleaning, grading, sorting, packaging, storage, transportation, and marketing of agricultural produce. These activities are essential for maintaining crops' quality, nutritional value, and marketability, thereby ensuring farmers' profitability. However, inadequate knowledge and lack of access to appropriate technologies and infrastructure often lead to post-harvest losses, reduced product quality, and limited market opportunities. Farm advisory services bridge this knowledge and technology gap by providing farmers with guidance and recommendations on post-harvest practices. Farm advisory services also support farmers in accessing market information and developing strong linkages within agricultural value chains. Advisors provide market intelligence, price trends, and information on consumer preferences, helping farmers make informed decisions about crop selection, production volumes, and marketing strategies. They also assist farmers in exploring value-added opportunities, diversifying products, and establishing market linkages with processors, retailers, and exporters, thereby enhancing their income potential and ensuring market sustainability.

Capacity Building and Training

Capacity building and training programs are essential components of farm advisory services. Advisors organize workshops, demonstrations, and farmer field schools to enhance farmers' knowledge and skills. Awareness and sensitization programms cover various topics, such as new farming techniques, technological advancements, good agricultural practices and business management. By equipping farmers with the necessary skills and knowledge, farm advisory services empower them to adopt best practices, improve their productivity, and enhance their overall farm management capabilities.

Monitoring and Evaluation

Monitoring and evaluation is essential to ensure farm advisory services' effectiveness. Advisors track farmers' progress, assess their interventions' impact, and collect data on key performance indicators. Monitoring and evaluation enable advisors to identify areas of improvement, measure the effectiveness of their recommendations, and tailor their services to meet the specific needs of farmers. Mobile based services plays significant impact in ensuring farm sustainability.

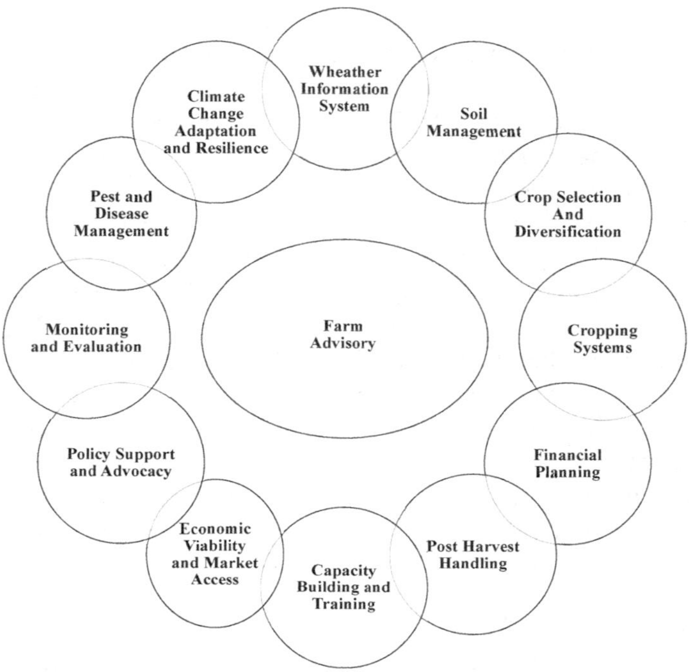

Fig. 3: Various Aspects of Farm Advisory

Resource Conservation in Saline Agroecological Systems

Efficient utilization of resources, such as water, soil, nutrients, and energy, is essential to optimize productivity, conserve resources, and minimize the environmental impact of agricultural practices (Adak and Babu, 2021). Effective resource management is crucial in mitigating the adverse effects of saline-sodic stress. Resource management practices tailored for saline-sodic ecosystems involve various approaches. For instance, implementing appropriate drainage systems, such as subsurface drainage or contour trenches, can help regulate the water table and flush excess salts from the soil profile. Additionally, the use of salt-tolerant crop varieties and improved irrigation methods, such as drip or sprinkler irrigation can aid in water conservation and minimize salt accumulation. Various strategies that can be used for resource conservation are presented in Fig. 4.

Water Conservation

- Use efficient irrigation methods such as drip irrigation or micro-sprinklers to minimize water wastage.
- Implement water-saving techniques such as mulching, which helps retain soil moisture and reduce evaporation.

- Collect and reuse rainwater through rainwater harvesting systems.
- Monitor soil moisture levels and adjust irrigation schedules accordingly to avoid overwatering.

Soil Conservation

- Practice soil erosion control measures such as contour ploughing, terracing, or windbreaks to prevent soil erosion.
- Use cover crops or green manure to protect the soil from erosion and enhance organic matter content.
- Implement conservation tillage or minimum tillage practices to reduce soil disturbance and improve soil structure.
- Apply organic matter, such as compost or farmyard manure, to enhance soil fertility and moisture retention.

Nutrient Management

- Conduct regular soil testing to determine the nutrient status of the soil and apply fertilizers accordingly.
- Going for slow-release or controlled-release fertilizers to minimize nutrient leaching and improve nutrient use efficiency.
- Utilize organic fertilizers and nutrient-rich compost to reduce reliance on synthetic fertilizers.
- Practice crop rotation to prevent nutrient imbalances and improve soil health.

Energy Conservation

- Utilize energy-efficient equipment, such as energy-saving irrigation pumps or energy-efficient lighting systems in greenhouses.
- Install renewable energy systems like solar panels to generate clean energy for farm operations.
- Optimize energy usage by scheduling farm activities during off-peak hours and implementing energy-saving practices in buildings and facilities.

Integrated Pest Management

- Implement integrated pest management (IPM) techniques to reduce the use of chemical pesticides.
- Encourage natural predators and beneficial insects to control pests.

- Use pheromone traps, sticky traps, or other non-chemical methods for pest monitoring.
- Employ cultural practices such as crop rotation, proper sanitation, and resistant crop varieties to minimize pest and disease pressure.

Waste Reduction and Recycling

- Implement a proper waste management system, including recycling, composting, and responsible disposal of farm waste.
- Reuse or recycle materials such as plastic containers or packaging wherever possible.
- Minimize food waste through proper harvesting, handling, and storage practices.
- Minimize packaging waste by using eco-friendly packaging materials or exploring alternatives like bulk sales or reusable containers.
- Implement recycling programs for materials such as plastic, cardboard, and other recyclable waste generated on the farm.
- Compost organic waste is generated from crop residues, trimmings, and spoiled produce to produce nutrient-rich compost for the farm.

Efficient Farm Planning

- Optimize farm layout and design to minimize resource wastage and improve operational efficiency.
- Consider factors such as proximity to water sources, efficient use of space, and minimizing travel distances within the farm.
- Regularly maintain and service farm machinery to ensure they operate at peak efficiency and consume less fuel.
- Going for energy-efficient equipment and machinery when making purchasing decisions.
- Properly calibrate and adjust equipment for accurate application of fertilizers, pesticides, and irrigation water.

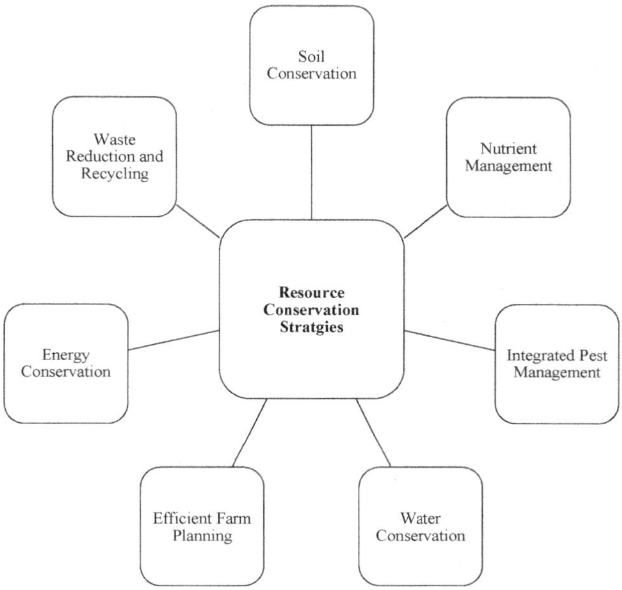

Fig. 4: Various Strategies of Resource Conservation

Farm advisory services play a pivotal role in supporting farmers in saline agroecological systems by providing specialized knowledge, robust recommendations, capacity building and policy support. By focusing on soil management, crop selection, efficient irrigation practices, integrated pest and disease management, and market access, farm advisors enable farmers to navigate the challenges posed by soil salinity and achieve sustainable agricultural productivity. Through collaboration with research institutions, policy advocacy, and continuous monitoring and evaluation, farm advisory services contribute to developing resilient and thriving productivity systems in saline agroecological environments. Challenges encountered in saline and sodic agroecosystems are multifaceted, encompassing soil-related, water-related, and plant-related factors. Overcoming these challenges requires comprehensive and integrated approaches focusing on improving soil fertility, enhancing water management, and selecting suitable crops. Continuous monitoring and evaluation systems enable advisors to identify areas of improvement, measure the effectiveness of their recommendations and improve services to meet the specific needs of farmers. Digital services were thus plays significant impact on farmers' income and attaining sustainability.

References

Adak, T., Babu, N., Pandey, G. and Kumar, A. (2020). Soil moisture and climate analysis for estimating stresses in orchards. In: National Web-Conference on "Sustainable Soil and Water Management for Bio-diversity Conservation, Food Security and Climate Resilience" held during 29 to 30th December 2020, p. 17.

Adak, T., Kumar, K. and Pandey, G. (2021). Assessing yield sustainability in guava fruit for greater remuneration to growers. In: 2nd Asian Web Conference "Managing Hill Resources and Diversities for Zero Hunger and Climate Resilience" held during 12-13 February, 2021. p 39.

Adak, T., Rajan, S. and Singh, V. K. (2018). Dynamics of Soil and Tree Carbon Storage in Different Agroforestry/Tree based Land Use Systems. Journal of Agricultural Physics 18 (1): 127-134.

Adak, T. and Babu, N. (2021). Soil and air permeability - A key factor for soil processes and productivity. Current Advances in Agricultural Sciences (An International Journal), 13(1): 13-18.

Adak, T. and Pandey, G. (2021). Assessment of soil biological activity as an ecological indicator for mango orchard soils. Journal of Natural Resource Conservation and Management 2(2): 133-138.

Babu, N., Adak, T., Chandra, S. and Kumar, A. (2020). Crop Diversification in Fruit Belts - Sole Way to Enhance Farmers Income. In: Soil and Water Conservation Bulletin 5, Dogra, P., Sharma, N. K., Mandal, D., Kumar, G., Rawat, I., Bishnoi, R., Kar, S. K. and Kumar, S. (Eds), Indian Association of Soil and Water Conservationists, Dehradun, Uttarakhand, 2020, pp 89-94.

Babu, N., Adak, T. and Kumar, A. (2022). Horticulture intervention for improving productivity and profitability of small and marginal farmers. Journal of Natural Resource Conservation and Management, 3(1): 49-57.

Banyal, R., Sanwal, S. K., Sharma, P. C., Yadav, R. K. and Dagar, J. C. (2019). Fruit and vegetable-based saline agricultural systems for nutritional and livelihood security. Res. Dev. Saline Agric. 729-751.

Barman, P., Kumar, D., Kumar, R., Mishra, D., Trivedi, A. K. and Pandey, A. K. (2021a). Holistic Pest Management in Mango Orchards by Manipulating Canopy Architecture: A Review. J Hortics p 374.

Barman, P., Kumar, R., Pandey, A., Bishnoi, C., Gora, J., Berwal, M., Dhaka, S., Sarolia, D., Kumar, M. and Pratibha, A. (2021b). Can canopy management increase quality fruit production in *Syzygium cumini* (L.) Skeels? Eur. J. Hortic. Sci 86: 371-383.

Dagar, J. C. (2013). Greening salty and waterlogged lands through agroforestry systems for livelihood security and better environment, in: Agroforestry Systems in India: Livelihood Security and Ecosystem Services. Springer, pp. 273-332.

Dagar, J. C., Yadav, R. K. and Sharma, P. C. (2019). Research developments in saline agriculture. Springer.

Gora, J. S., Singh, V. K., Sarolia, D. K., Kumar, K., Rajkumar and Bhati, V. (2017). Performance of mango (*Mangifera indica* L.) monoembryonic and polyembryonic seedlings under salt stress condition. Int. J. Curr. Microbiol. Appl. Sci. 6: 3051-3056.

Gundappa, Adak, T. and Shukla, P. K. (2016a). Seasonal dynamics of mango hoppers and their management under subtropics. GERF Bull Biosc, 7(1): 6-9.

Gundappa, Adak, T. and Shukla, P. K. (2016b). Appraisal of thrips population dynamics in mango using weather based indices. Vegetos, 29(3): 138-145.

Gundappa, Adak, T. and Shukla, P. K. (2016c). Seasonal incidence and weather based forecasting model for mango leaf webber under subtropical conditions. Current Advances in Agricultural Sciences, 8(1): 96-99.

Gupta, S. R. and Dagar, J. C. (2016). Agroforestry for ecological restoration of salt-affected lands. Innov. saline Agric. 161-182.

Haj-Amor, Z., Araya, T., Kim, D. -G., Bouri, S., Lee, J., Ghiloufi, W., Yang, Y., Kang, H., Jhariya, M. K. and Banerjee, A. (2022). Soil salinity and its associated effects on soil microorganisms, greenhouse gas emissions, crop yield, biodiversity and desertification: A review. Sci. Total Environ. 843: 156946.

Jinger, D., Kaushal, R., Kumar, R., Paramesh, V., Verma, A., Shukla, M., Chavan, S. B., Kakade, V., Dobhal, S. and Uthappa, A. R. (2023). Degraded land rehabilitation through agroforestry in India: Achievements, current understanding, and future prospectives. Front. Ecol. Evol 11: 69.

Kumar, A., Mehla, M. K. and Kumar, S. (2020). A Review: Use of Saline Water in Irrigation for Agriculture. Ind. J. Pure App. Biosci 8: 262-271.

Mehla, M. K., Jhorar, R. K., Kumar, S. and Kumar, A. (2022). Effect of Drip Irrigation Frequency and Lateral Spacing on Kharif Onion Crop. Int. J. Ag. Env. Biotech 15: 205-212.

Rai, A. K., Basak, N. and Sundha, P. (2020). Chemistry of salt-affected soils, in: Minhas, P. S., Yadav, R. . K., Sharma, P. C. (Eds.), Salinity Management in India. Indian Council of Agricultural Research (ICAR), New Delhi, India.

Rajkumar, Singh, A., Mann, A. and Yadav, R. K. (2020). Performance evaluation of pomegranate (*Punica granatum*) genotypes under saline conditions. Indian J. Agric. Sci. 90: 117-122.

Sanwal, S. K., Kaur, G., Mann, A., Rajkumar and Rai, A.K. (2019). Response of okra (*Abelmoschus esculentus* L.) genotypes to salinity stress in relation to seedling stage. J. Soil Salin. Water Qual. 11: 217–223.

Sanwal, S. K., Mann, A., Kesh, H., Kaur, G., Kumar, R. and Rai, A. K. (2021). Genotype environment interaction analysis for fruit yield in okra (*Abelmoschus esculentus* L.) under alkaline environments. Indian J. Genet. Plant Breed. 81: 101–110.

Sethi, M., Bundela, D. S. and Rajkumar (2016). Diagnosis and prognosis of salt-affected soils and poor-quality waters using remote sensing and proximal techniques. Innov. Saline Agric. 55-82.

Sharma, P. C. and Singh, A. (2019). Reviving the productivity of salt-affected lands: Technological options, constraints and research needs. Res. Dev. Saline Agric. 591-627.

Shukla, P. K., Gundappa and Adak, T. (2017). Development of sooty moulds in mango orchards in relation to weather parameters and major sucking pests. Journal of Environmental Biology, 38(6): 1293-1300.

Singh, A., Kumar, A., Sharma, P. C., Kumar, R. and Yadav, R. K. (2023). Sodicity stress differently influences physiological traits and anti-oxidant enzymes in pear and peach cultivars. Peer J 11: e14947.

Singh, A., Sharma, D. K., Kumar, R., Kumar, A., Yadav, R. K. and Gupta, S. K. (2018). Soil salinity management in fruit crops: a review of options and challenges. Eng. Pract. Manag. Soil Salin. Agric. Physiol. Adapt. Approaches 39-85.

Singh, A., Singh, R. K., Kumar, Arvind, Kumar, Ashwani, Kumar, R., Kumar, N., Sheoran, P., Yadav, R. K. and Sharma, D. K. (2021). Adaptation to social-ecological stressors: a case study with Indian jujube (*Ziziphus mauritiana* L.) growers of north-western India. Environ. Dev. Sustain. 23: 3265-3288.

Singh, G. (2009). Salinity related desertification and management strategies: Indian experience. L. Degrad. Dev. 20: 367-385.

Tuomisto, H. L., Scheelbeek, P. F. D., Chalabi, Z., Green, R., Smith, R. D., Haines, A. and Dangour, A. D. (2017). Effects of environmental change on agriculture, nutrition and health: A framework with a focus on fruits and vegetables. Wellcome open Res. 2.

14

Agro-Advisory Services and Their Implication for Sustainable Horticultural Production

Yogesh Kumar[1], Ved Prakash[2]* and Dharmendra Singh[3] and Ashutosh Upadhyaya[5]

[1]CCSHAU College of Agriculture, Bawal, Rewari, Haryana
[2,5]Division of Land and Water Management, ICAR- Research Complex for Eastern Region, Bihar
[3]Department of Extension Education, CCS Haryana Agricultural University Hisar, Haryana

Introduction

Horticulture is distinct from other economic sectors as its output is heavily reliant on weather conditions. The degree of success of horticultural production and economics is decided to a large extent by how well weather conditions corresponding to the crop's optimal requirements are best utilized to raise the crops. Also, how successfully adverse weather circumstances, such as moisture, thermal, wind, radiation and biotic stresses were handled to minimize adversity. It is well understood that variability in weather elements (particularly rainfall and temperature) is the primary cause of inter-annual variability in crop output. The onset of weather extremes (dry spells, droughts, floods, heat waves, and hailstorms) has a negative impact on horticultural yield, resulting in a low level of productivity. Furthermore, weather can have an impact on output at multiple levels (vegetative stage, harvest, transportation, storage, *etc.*). Flooding, droughts, heat waves, cold spells, hailstorms, and other natural calamities pose significant risks to farms. These natural catastrophes can lead to lower yields, leaving uninsured farmers with little income for the season. Farmers often plant low-risk, low-return crops (such as early maturing varieties that are drought-resistant) to deal with erratic weather rather than investing in more profitable crops that are more sensitive to weather fluctuations and extremes. Thus, atmospheric stress has significant impacts on the food ecosystem (Bal and Minhas, 2017; MANAGE, 2018).

Many risks connected with farm output are influenced by a region's climate and seasonal weather, which ultimately determines farm income. As a result, farmers may become trapped in a cycle of low productivity as a consequence of extreme weather events. This necessitates an assessment of the vulnerability of any location or area. Climate change's effect on Indian agriculture and water resources is a major concern for policymakers and development experts. Agricultural sector in India is extremely sensitive to changes in rainfall patterns. Therefore precise water management was advised to fruit growers (Adak and Babu, 2022). Horticultural technologies were disseminated to small and marginal farmers to improve farm productivity and profitability (Babu *et al.*, 2021). Since disease and insects significantly affects the farm productivity, their management is very crucial under changing weather condition (Shukla *et al.*, 2016; Gundappa *et al.*, 2015). The role of nutritional advisories to growers like boron management is essential for improving quality of fruits (Shukla *et al.*, 2020) while Adak *et al.* (2022) advocated spraying zinc for improving yield and pulp nutrient contents in Dashehari mango. Because climate change affects the nation's food security, the magnitude of the shift and its consequences will be of concern to scientists, planners, and policymakers, particularly in terms of regulating the food supply and preserving reserves for the future. Meteorological conditions have the greatest impact on high-yielding agricultural varieties that are more sensitive to environmental conditions, necessitating maximal optimization of water, air, thermal, and nutritional conditions. As a result, the importance of agro-meteorological knowledge is growing. Using educated information on the impact of weather and climatic factors on horticultural productivity can not only reduce damage, but also allow for additional yield without major financial orchard operations. Weather forecasts in all temporal ranges are preferable for effective planning and administration of agri-horticultural practices. Weather forecasts in the short and medium ranges greatly assist in making short-term changes in daily operations that minimize losses caused by adverse weather conditions and improve yield, quantity, and quality of horticultural productions.

The Major Types of Climate Information Needed by Farmers Before and During Crop Production are as Follows (FAO 2019)

- Before the vegetable cropping season begins, an agrometeorological crop-risk analysis is performed to determine the most appropriate crop(s) based on crop water requirements and forecasted seasonal rainfall.
- Determination of the best planting date for a particular region based on rainfall onset.

- Seasonal climate outlooks, which can be used to prepare farmers for a variety of scenarios that, may emerge during the season as a result of uncertainty in long-term weather forecasts. Crop including vegetable contingency plans can be developed for a variety of situations.
- Three-day weather forecasts throughout the growing season, with a focus on the likelihood of extreme weather events; 10-day crop management advisories based on the forecasted weather.

Agromet Advisory Service Scheme in India

The Agrometeorological Advisory Service (AAS) provided by India Meteorological Department (IMD), Ministry of Earth Sciences (MoES), is a vibrant step towards contributing to weather-based crop/livestock management strategies and operations aimed at increasing crop output and food security. The current AAS system's primarily focused on collection and organize climate/weather, soil, and crop information, and to combine it with weather forecast to help farmers in making management decisions. This has aided in the development and implementation of operational tools to handle weather-related uncertainties for efficient horticulture in rapidly changing environments via agro-meteorological applications.

Milestones in Agromet Services

- 1932 – Agromet services
- 1945 - Farmers Weather Bulletin
- 1976, IMD initiated Agromet Advisory Services at the state level in collaboration with the agriculture departments of the respective state governments.
- In 1991, the National Centre for Medium Range Weather Forecasting (NCMRWF) started preparing quantitative medium-range weather forecasts for Agromet Field Units (AMFUs). The expansion to 134 agro-climatic zones occurred later, specifically in 2008.
- 2008 - AAS at District Level (636 Districts)
- 2012 - Gramin Krishi Mausam Sewa (GKMS)
- 2018 - Experimental AAS at Block Level (200 Blocks in 24 States) and plans to extend to ~ 6500 blocks.

Timeline of Agro-met Advisory Dissemination Mechanism in India

- **From 1996 to 2007**, advisory bulletins were mailed to progressive farmers, the mass media (AIR, newspapers, and TV channels), the KVK, the State Department of Agriculture and Horticulture, non-governmental organisations (NGOs), field and horticulture crop growers associations, and poultry units for wider distribution.

- **From 2008 to 2010**, the reports were emailed to all of the above stakeholders except farmers. The bulletins were also distributed to Agrimet Division Pune, Reliance, Ruiter, IFFCO, I-kisan, and Nokia for further distribution to farmers and other end consumers.
- **From 2010 to 2015,** the bulletins were uploaded to the IMD website, AICRPAM (Crop weather outlook), and university websites in addition to the above means of distribution.
- **Since 2015,** Bulletins have been prepared online in the Farmers Portal, which was introduced by the Ministry of Agriculture, Government of India.

AAS provide very special inputs to farmers in the form of advisories that can make a huge difference in agriculture output by taking advantage of benevolent weather and minimizing the negative impact of malevolent weather. This has the ability to transform India's face in terms of food security and poverty alleviation. In India, IMD, MoES, is operating an operational AAS, Gramin Krishi Mausam Sewa (GKMS), at the district level, which represents a small step towards agri-horticultural management that is in tune with weather and climate variability, leading to weather proofing for farm output. AAS defines the demands of the farming community by determining the information requirements of various groups of end-users. The primary need of the farmer has evolved as location specific weather forecasting in quantitative terms. As a result, it was created and made active in June of 2008. Following that, a mechanism was created to integrate weather forecast and climatic information with agro-meteorological information in order to produce district level agromet advisories outlining farm management actions to capitalize on favourable weather and mitigate the effects of adverse weather (Mall *et al.,* 2006; Chukwuji *et al.,* 2019). To improve information dissemination, a system for communicating and disseminating agro-meteorological advisories has been created. For resource poor farmers', Kumar *et al.* (2016) suggested scientific intervention in sustaining mango productivity while Shukla *et al.* (2016) advised farmers for efficient shoulder browning disease management in mango. To enhance resources use efficiency, drip fertigation systems were highly recommended to farmers (Adak *et al.,* 2018). Web based advisory services were thus played significant role in diffusion of farmers' friendly technologies *vis-à-vis* improving the economic status.

Application of ICT to Address Climate Change Issues

There is much scope for technological developments to reduce GHG emissions in the agri-horticultural and livestock sector. For example, increases in crop and animal productivity will be implemented through improved management and husbandry techniques, such as better management, genetically modified crops, improved cultivars, fertilizer recommendation systems, precision farming, improved animal breeds, shelter management, improved animal nutrition, dietary additives and growth promoters, improved animal fertility, bio-energy crops, anaerobic slurry digestion manure disposal mechanisms and methane capture systems. And also in these line ICT could deliver promptly.

1. Awareness creation about climate resilient horticultural practices.
2. Sensitization of line dept. officials and field functionaries about the adverse climate effect and adaptation practices.
3. Preparedness towards the external climate impact and implementation of appropriate measures.
4. Evolving the risk aversion strategies with multi-stakeholders.
5. Framing the cope-up methods for resilient practices and reduce the vulnerability.
6. Awareness on carbon trading and conservation horticulture.

For site specific weather forecast and Agromet Advisory Billetin Farmers can download and use some significant weather app

- "MAUSAM App" for location-specific forecast and warning: (https:// play. google. com/store/apps/details?id=com. imd. masuam)
- "Meghdoot App" for Agromet advisory (https://play. google. com/store/ apps/details?id=com. aas. meghdoot)
- "Damini App" for Lightning warning: Damini Lightning apps is developed by IITM-Pune and ESSO. The apps is monitoring all lightning activity which are happening in specifically for all india. This app provides alert, if lightning is happening near you by GPS notification. under 20 KM and 40 KM. (https://play. google. com/store/ apps/details?id=com. lightening. live. damini)
- Public Observation – IMD for Application provides weather information to users location wise. Users can also provide their feedback on this app. (https://play. google. com/store/apps/details?id=com. mausam. crowdsource)
- "Umang App" for UMANG (Unified Mobile Application for New-age

Governance) is envisaged to make e-governance 'mobile first'. It is developed by Ministry of Electronics and Information Technology (MeitY) and National e-Governance Division (NeGD).

- (https://play. google. com/store/apps/details?id=in. gov. umang. negd. g2c)

The following Climate-Extension-Program framework will also depict the climate impact program at field level.

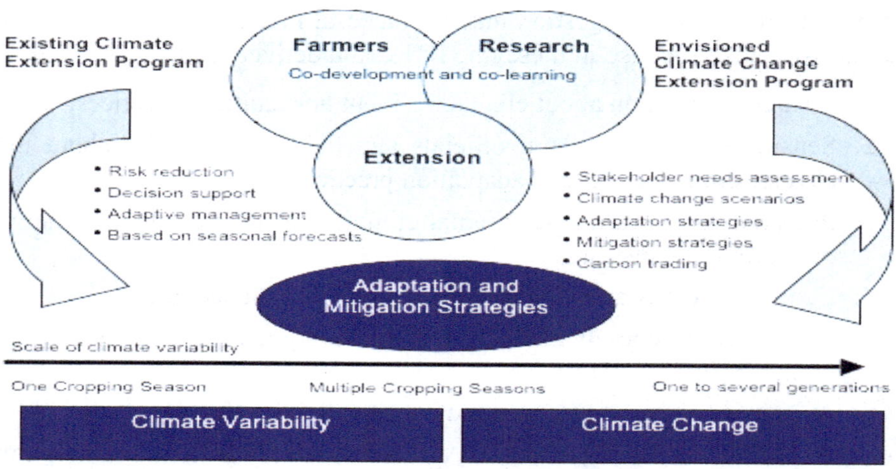

Fig. 1: Framework of a Combined Climate Variability and Change Extension Program (Adapted from Fraisse *et al.*, 2009)

Disseminating Information Through Extension in Light of CSA

Agromet advisories are distributed to farmers via different multi-channel systems such as All India Radio and Doordarshan, private TV and radio channels, newspapers, the internet, and SMS. The National Bank for Agricultural and Rural Development, for example, sends out agromet warnings to farmers via SMS and IVR. In addition, in collaboration with the National Informatics Centre (NIC)/Agricultural Technology Management Agency (ATMA)/KVK/NABARD/Internet, a number of AMFUs have begun issuing agromet advisories via SMS. Agromet Advisories are also being disseminated in both regional and English languages via kisan SMS, a portal established by the Government of India's Ministry of Agriculture and Farmers Welfare. Currently, this service immediately benefits 21.69 million farmers (Chattopadhyay and Chandras, 2018). To foster higher yields and income from per unit land area, high density planting systems were also recommended to farmers (Adak *et al.,* 2019).

Extension Approaches Used in CSA

Addressing these global challenges necessitates generational adaptation and the application of new knowledge, which necessitates interaction and support from a diverse variety of organizations. Extension systems address a wide range of goals that include and extend far beyond the transmission of new technology. This includes responding effectively and responsively to domestic and foreign markets in order to reduce vulnerability and strengthen the voice of the rural poor. Extension systems can add to CSA in a variety of ways. The provision of agro meteorological services by the India Meteorological Department (IMD), Ministry of Earth Sciences, is a step towards contributing to weather-informed management strategies and operations aimed at increasing crop output and food security. IMD is running the Gramin Krishi Mausam Sewa (GKMS) project, which aims to service the farming community in various parts of the nation and has the potential to change the face of any country in terms of food security and poverty alleviation.

Innovative Extension Approaches for Climate Smart System

Short Messaging Services

Farmers' registered mobile numbers received short text messages of 160-164 characters in the local tongue (Hindi, English, Marathi and local language in that specified areas). Farmers received no more than two SMS messages per week based on weather warnings and contingency plans.

Climate Wallpapers

To provide advisory services to farmers, one-page advisories in the shape of tables or posters related to weather prediction of agricultural operations required to be done were pasted on common display boards of villages. These advisors were dubbed Krishi Salah in the CCA initiative.

Climate Voice Messages

SMS was converted into voice messages and distributed to farmers in regions with poor literacy rates.

Climate Group Meetings

Farmers were organized in groups because it is easier to connect and spread climate-related information to them in groups. Various committees were established in the village to oversee various components, such as the CCA project's village development committee (VDC). A committee had 10-15 members, and there was a 40-50% allocation for women in all committees (MANAGE, 2018).

Economic Assessment of the Agromet Advisory Services

According to a study conducted by the National Centre for Agricultural Economics and Policy Research in 1996 to evaluate the economic effect of weather forecast-based advisories in 1996, 2009, and 2015, farmers received a 10-25% economic benefit from the use of agromet advisory services. The National Council of Applied Economic Research estimated the economic benefit of these services in 2009 at Rs. 50,000 crores per year, which could be increased to Rs. 211,000 crores if the entire farming community in the nation used Agromet information in their agricultural activities. In 2015, the National Council of Applied Economic Research estimated the economic benefit of weather information use as a product of the proportion of farmers getting information. Crop-specific conversion factors were used to convert farmers' financial earnings to economic profits. At the moment, only 24 percent of farms are taking advantage of SMS services. When all farming households use Agromet Services, it has the ability to generate a net economic benefit of up to Rs. 3.3 lakh crores on the 22 major crops (Chattopadhyay and Chandras, 2018).

Key Benefits from Weather Forecasting to Different Sectors

The availability of timely weather information allows farmers to organize their operations in such a way that not only costs and crop losses are minimized, but yield gains are maximized. These weather forecasts aid in better orchard management for fruits, vegetables and flower cultivation. Small farmers with small cultivable land whose agricultural production is dependent on the monsoon are the largest beneficiaries. Improved weather forecasts are being used by economic sectors all over the world to optimize their activities. According to the World Bank, the key sectors expected to benefit the most out of the 30 or so are:

Agriculture and Related Industries (Farm and Fisheries): The agriculture industry could benefit greatly for the following reasons:

- Accurate predictions would allow for timely agricultural sowing, ploughing, irrigation, and harvesting. Increased precision would help with fertilizers application timing, pest and disease management, and preventing over-applications that are washed or blown away.
- Forecasting could direct the cost-effective use of special preventive measures to mitigate the damage caused by wind and precipitation such as frost, hail, drought, and erosion.
- Forecasting could be used to optimize agricultural product storage and transportation, reducing post-harvest losses.

Dissemination of Agro-meteorological Bulletins

AAS's mission is to provide knowledge that will assist farmers in making the best use of weather and climate resources. Although concerted efforts are being made to establish two-way communications, the information flow is currently mainly one-way. Because agro-meteorologists at Agro-meteorological Field Units (AMFUs) interact with farmers less frequently, good communication and working relationships have been established with agricultural extension personnel such as Krishi Vigyan Kendras, Kisan Call Centres, and others to promote participatory methods for interactions with farmers. Due consideration is being given to the message's content, which is pertinent to the farmer's upcoming weather-based decision making. This includes identifying weather and climate sensitive choices and interacting with weather forecasters from IMD Meteorological Centres and agriculture specialists from SAUs and ICAR. Under the Public Private Partnership mode, AMFUs disseminate supportive agromet advisories to farmers via various multi-channel systems such as All India Radio (AIR) and Doordarshan, Private TV and radio networks, Newspaper and Internet, SMS and IVRS. Agromet warnings are also distributed through the Ministry of Agriculture's Kisan Portal.

http://farmer. gov. in

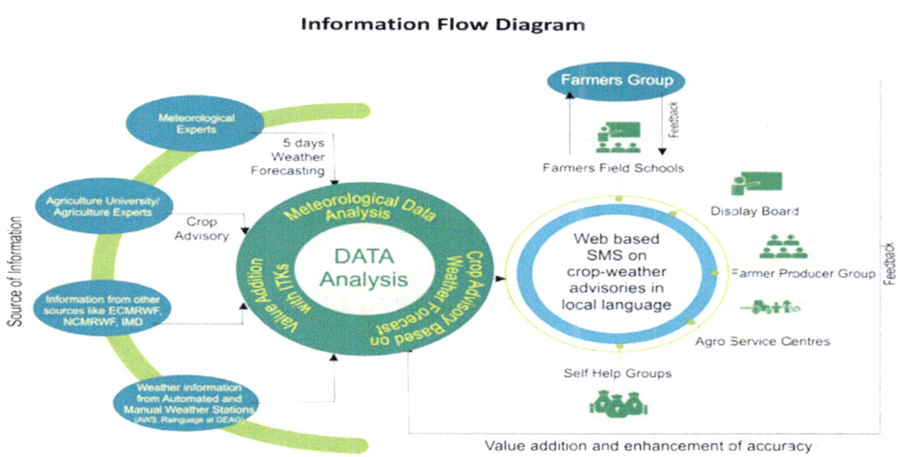

Fig. 2: Framework for flow diagram for data generation in web based advisories like Agro-DSS portal

Essential Components of Agrometeorological Advisory are

- Weather forecast.
- Warning of extreme weather events.

- Information related to location-specific normal sowing practices of crops.
- Farm management details, such as irrigation scheduling, insect and disease control spraying, nutrient (e.g., N, P, K and micronutrients) application, and so on.
- Suggested means for forecasted weather for farmers to take proper action based on their situation.
- Special alert for pest and disease outbreaks, as well as suitable control measures.
- Advice on livestock administration and animal husbandry.

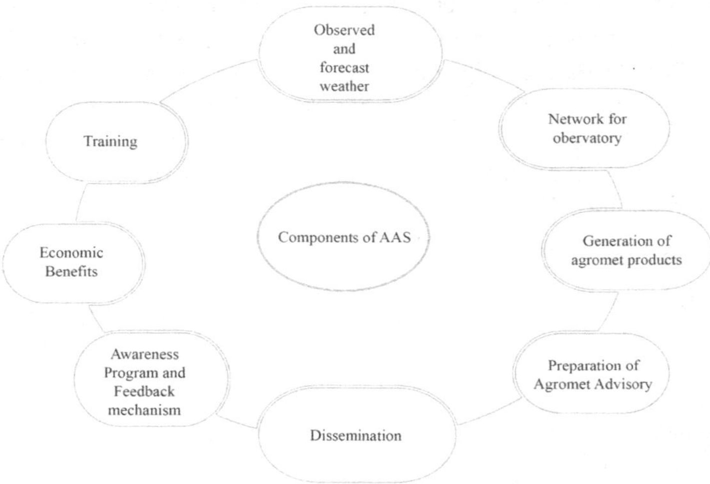

Fig. 3: Flow chart of a combined component of Agro-Advisory Services for better utility of AAS

Role of AAS in Improving Production and Ensuring Food Security

Yield sensitivity to seasonal variations in climate is a significant cause for inter-year crop production. Climate variability has increased in the nation, both spatially and temporally. The number of days with no rain (dry spells) and heavy rain is rising (Singh and Malik, 2018) and a well-distributed monsoon season has become increasingly uncommon in recent years. Farmers were advocated to grow turmeric in the interspaces of orchards (Shukla and Adak, 2020) and also to enhance farmers income, diversification with vegetables, flowers, spices and other medicinal and aromatic crops can be tried as well (Babu *et al.*, 2020). A timely agromet advisory can help to minimize crop loss. For example, real time based advisory can saved anthracnose disease problems and mango mealybug in relation with weather parameters (Shukla *et al.*, 2017; Gundappa *et al.*, 2018). Drip fertigation systems were advocated to farmers for enhancing water use efficiency and productivity (Adak *et al.*, 2015). Vashisth *et al.* (2013) also emphasied production and income of the farmers under changing climate scenario. Another element is the dissemination of scientific cultivation practices in accordance with weather forecasts to farmers via AAS. This will ensure that farmers complete all agronomic and crop protection activities on time, resulting in increased crop output.

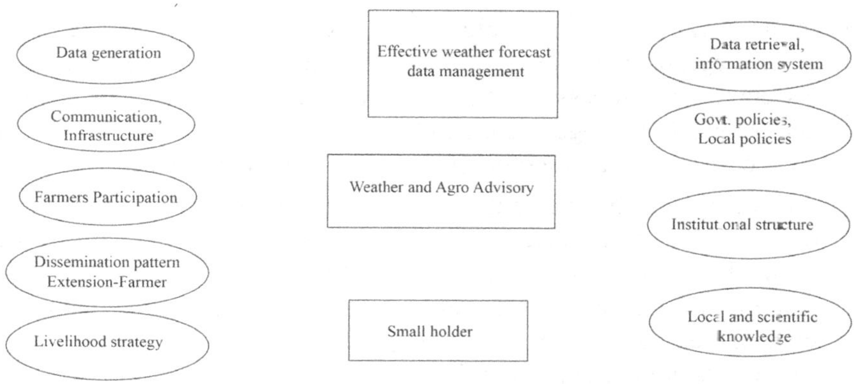

Fig. 4: Flow chart of a combined component of AAS and various stakeholders for better utility of AAS

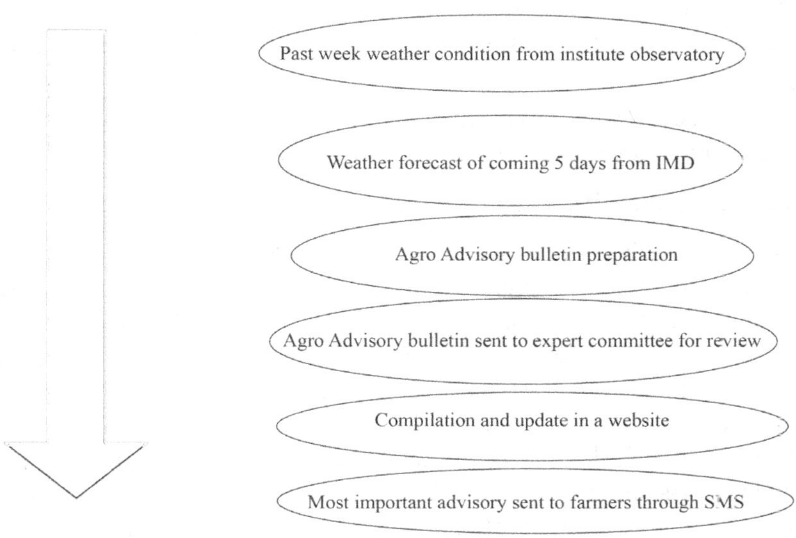

Fig. 5: Flow chart of a region specific generation of Agro-Advisory Services at micro-level.

Opportunities Ahead and Policy Implications

As micro-scale, advisories become more important; let us expect that the spatial resolution and accuracy of weather forecasts will improve as computing skills increase over time.

- Developing a dedicated platform for high-quality weather and climate data from both public and commercial institutions/agencies. The Ministry of Earth Sciences and the Ministry of Agriculture and Farmers' Welfare of the Government of India must work together on this.
- Implementing AAS using block-level weather forecast for areas with diverse agro-ecologies will necessitate automation of development and dissemination processes with minimal human using machine learning and artificial language. Though IMD has already taken the lead in this area, collaboration with other partners must be strengthened.
- Enhancing capacity in agrometeorology, an interdisciplinary topic, should be expanded at all agricultural universities. Agricultural meteorology should be required as part of the doctoral curriculum. There is also great potential for the development of 'extension agrometeorology' as a full-fledged method for AAS awareness generation, advisory dissemination, and feedback gathering from farmers.
- The use of ICTs for AAS dissemination may be investigated further in order to increase the number of farmers addressed. Exploring the use of ICTs and involving other stakeholders (public and private institutions, farmer groups, and farmers) should be undertaken to improve the AAS overall.
- Farmers' traditional knowledge and practical experiences should be recorded and incorporated into the development of AAS.

Constraints Involved with AAS

The primary bottleneck in the dissemination of AAS is the lack of linkage between institutional players due to deficiencies in quality weather data availability, data sharing, coordination, and communication. The accuracy of the medium-term forecast (lead time of 5 days) is the main determinant of AAS success, as the accuracy of seasonal forecasts (lead time of 1 month) is presently doubtful, and is intended for policymakers rather than farmers. Another cause is the availability of dedicated human resources for advisory preparation and dissemination. There has also been no attempt to engage farmers in improving the content of AAS. This results in the development of AAS without regard for the needs and goals of farmers. Other causes include ineffective research programmes and insufficient use of information and communication technologies (ICTs), which leads to a low rate of AAS adoption (Singh *et al.*, 2019). AAS content should be very location specific, focusing on specific crops produced, market facilities, and so on. Initiative of regular basis AAS as text messages may become most effective method of reaching a large number of farmers across the country. As a result, AAS providers should devise methods to include them in the information dissemination cycle as well.

Future Challenges

- Extreme weather events such as heavy rain, cloud bursts, and hailstorms are causing extensive crop damage in many parts of India. Despite the fact that IMD issues 'nowcast bulletins' and 'special weather bulletins,' forecast accuracy needs to be better.
- With the advent of ICTs for AAS dissemination, the agency has been able to provide agencies to serve large numbers of farmers. However, many rural farms still lack access to mobile phones and the Internet.
- It has been observed that the same farmer receives different AAS for the same time period from different AAS providing organizations, which causes them confusion.
- Although the IMD issues block-level weather forecasts, AAS based on those forecasts is not upscaled to the national level.
- The final goal of AAS will be customized advisories for farmers. This will necessitate weather forecasts with farm-level spatial precision, farmer-level crop and soil intelligence, and massive computing power for AAS automation.

In this age of climate change, the significance of agromet advisory services is growing. Extreme weather events are becoming more common, presenting a significant threat to global crop production sustainability. This chapter has gone into great detail about weather-related risks, information required, the current system of AAS creation and dissemination in India, their economic effect, constraints, and future challenges. The twenty-first century has seen a significant increase in computing capacity, and thus in the dissemination of AAS via ICTs. Despite these advances, if we view the country as a hole, the dissemination of accurate AAS at the right moment to farmers is still lacking. The spatial resolution of the weather forecast has improved significantly, which has yet to be exploited in the shape of block-level AAS. Climate variability and change pose new threats and challenges to the current structure of AAS development and dissemination. Automation of AAS, further exploration of extension agrometeorology, and other opportunities are suggested for the immediate future. Issuance of time bound based advisories through mobile and whatsapp systems could revolutionize the productivity and profitability scenario.

References

Adak, T., Kumar, K., Singh, V. K. and Ganeshamurthy, A. N. (2022). Response of Dashehari mango to different Zn levels on yield and pulp nutrient contents grown on sandy loam soils of Lucknow. Journal of Horticultural Sciences 17(1): 103-109.

Adak, T., Kumar, K. and Pandey, G. (2018). Enhancing sustainable yield index in fertigated Mango cv. Dashehari. GERF Bulletin of Biosciences, 9(2):1-6.

Adak, T., Kumar, K. and Singh, V. K. (2015). Assessment of variations in reference evapotranspiration, yield and water use efficiency of mango under different fertigation regimes. Journal of Soil and Water Conservation 14(3): 232-240.

Adak, T., Mishra, D., Kumar, K. and Singh, V. K. (2019). Analysis of soil and tree productivity under high density planting system in mango cv. Dashehari (*Mangifera indica* L.). Tropical Plant Research 6(1): 24-30.

Adak, T. and Babu, N. (2022). Innovative approach for enhancing water productivity in Mango cv Amrapali. In: Proceedings of 3rd International Web-Conference on "Natural Resource Management for Global Food Security and Sustainable Development Goals" organized during December 2 to 3, 2022. p. 69-70.

Babu, N., Adak, T., Chandra, S. and Kumar, A. (2020). Crop Diversification in Fruit Belts - Sole Way to Enhance Farmers Income. In: Soil and Water Conservation Bulletin 5, Dogra, P., Sharma, N. K., Mandal, D., Kumar, G., Rawat, I., Bishnoi, R., Kar, S. K. and Kumar, S. (Eds), Indian Association of Soil and Water Conservationists, Dehradun, Uttarakhand, 2020, pp 89-94.

Babu, N., Adak, T. and Kumar, A. (2021). Some horticultural technologies for improving farm productivity and profit earning of small and marginal farmers under subtropical area. In: 9th Indian Horticultural Science Congress held during 18th to 21st November, 2021. p. 181.

Bal, S. K and Minhas, P. S. (2017). Atmospheric stressors: challenges and coping strategies. In: Minhas PS et al (eds) Abiotic stress management for resilient agriculture. Springer, Singapore, pp 9–50. https://doi. org/10. 1007/978-981-10-5744-1_2.

Chattopadhyay, N. and Chandras, S. (2018). Agro meteorological advisory services for sustainable development in Indian agriculture, Biodiversity International Journal. 2(1): 13-18.

Chukwuji, C. N., Gadanga, T. A., Sayudi, S., Yusuf, Z. and Zakariya, J. (2019). "Awareness Access and Utilization of Information on Climate Change by Farmers in Zamfara State, Nigeria", University of Nebraska - Lincoln Digital Commons @University of Nebraska - Lincoln, 1-25.

FAO (2019). Handbook on climate information for farming communities – what farmers need and what is available. FAO, Rome, p 184.

Fraisse, C. W., Breuer, N. E., Zierden, D. and Ingram, K. T. (2009). From Climate Variability to Climate Change: Challenges and Opportunities to Extension 25. April 2009. 47: 2-1-10 pp.

Gundappa, Adak, T. and Rajkumar, B. (2015). Variability of thermal indices and peak mango hopper incidence under subtropical conditions. Journal of Agricultural Physics 15(2):150-156.

Gundappa, Adak, T. and Shukla, P. K. (2018). Population dynamics of mango mealybug (*Drosicha mangiferae*) (Margorididae: Hemiptera) and its relation with weather parameters in subtropical climatic conditions. Indian Journal of Agricultural Sciences 88 (6): 865–870.

Kumar, K., Adak, T., Lal, B., Shukla, P. K., Balaji Rajkumar, Chandra, S. and, Singh, V. K. (2016). Technological intervention in sustaining mango productivity in a resource poor farmers' field of Uttar Pradesh. GERF Bulletin of Biosciences 7(2):6-11.

Mall, R. K., Singh, R., Gupta, A., Srinivasan, G. and Rathore, L. S. (2006). Impact of climate change on Indian agriculture: A review', Climatic Change, (2006) 78: 445-478, Springer Science, Business Media.

MANAGE, (2018). "Climate Smart Agriculture and Advisory Services: Approaches and Implications for Future", National Institute of Agricultural Extension Management. An organization of Ministry of Agriculture and Farmers' Welfare, Govt. of India. Rajendranagar, Hyderabad - 500 030, Telangana State, India.

National Innovations on Climate Resilient Agriculture—NICRA (2018) Research highlights 201618. ICAR-Central Research Institute for Dryland Agriculture, Hyderabad. p. 128

Shukla, P. K., Adak, T., Misra, A. K. and Singh, A. (2016). Appraisal of shoulder browning disease of mango (*Mangifera indica* L.) caused by Fungus complex in subtropical region of India. Journal of Mycology in Plant Pathology 46(1): 38-46.

Shukla, P. K., Adak, T. and Gundappa (2017). Anthracnose disease dynamics of mango orchards in relation to humid thermal ratio under subtropical climatic condition. Journal of Agrometeorology 19(1): 56-61.

Shukla, P. K., Adak, T. and Gundappa. (2016). Seasonal dynamics of powdery mildew of mango and its management under subtropics. GERF Bull Biosc 7(1):21-25.

Shukla, S. K., Adak, T. and Singh, V. K. (2020). Evaluation of boron nutrition in enhancing productivity and quality of mango (*Mangifera indica* L.) cultivar Mallika under subtropical climatic conditions. Current Advances in Agricultural Sciences 12(1): 33-36.

Shukla, S. K. and Adak, T. (2020). Intercropping of turmeric for revenue generation from interspaces in mango orchards. Kahaar 7(1-2): 41-43.

Singh, M., Ghanghas, B. S., Sharma, V. and Sharma, B. C. (2019). Minimize weather risk in agricultural planning and management through agromet advisory services in rural areas. In: Ahamad S, Sharma JP (eds) Transformation of Indian agriculture through innovative technologies. Daya Publishing House, Delhi, India, pp 11–21.

Singh, M. and Malik, J. S. (2018). Role of extension personal for transfer of information on climate change in agriculture: A case study of Haryana, India, Agricultural Science Digest 38(3):197-200.

Vashisth, A., Singh, R., Das, D. K. and Baloda, R. (2013). Weather based Agromet advisories for enhancing the production and income of the farmers under changing climate scenario. International Journal of Food Science and Technology 4(9):847–850.

15

Role of Farm Women in Sustainable Production of Horticulture Crops

***Ankita Sahu*[1] *and Monika Patel*[2]**

[1]*ICAR-Central Institute for Women in Agriculture, Bhubaneswar, Odisha*
[2]*Scientist (Floriculture), Horticulture College, Khuntpani, Birsa Agricultural University, Ranchi, Jharkhand*

Introduction

Horticulture crops hold significant importance for their economic, nutritional, environmental, and aesthetic contributions. They contribute to agricultural GDP, create employment opportunities, and provide income for farmers and other stakeholders along the value chain. Horticultural crops often have high-value market demand, leading to increased profitability for farmers and rural communities. The horticultural crops, including fruits, vegetables, nuts, and medicinal, spices are rich sources of essential nutrients, vitamins, minerals and dietary fiber. They contribute to a balanced and nutritious diet, promoting good health and preventing various diseases. Consumption of horticultural crops is linked to reduced risks of chronic diseases, including heart disease and certain types of cancer. These crops have environmental benefits. They contribute to biodiversity conservation by providing habitats for various species and supporting beneficial insects and pollinators. Additionally, horticultural practices such as agroforestry, crop diversification, and organic farming methods can enhance soil health, reduce erosion, promote water conservation, and mitigate climate change impacts. Adak and Pandey (2018) emphasised soil management of horticultural crops for attaining food and nutritional security of farm women. Babu *et al.* (2022) explained scientific interventions for enhancing productivity and profitability of farmers' engaged in horticultural sector while Gundappa *et al.* (2016c) recommended management protocols for effective controls of mango hoppers to growers. Even, digital solutions are now-a-days is being tried to improve economic prosperity of fruit growers (Verma and Adak, 2021). Horticulture crops contribute to food security by diversifying the food supply. They provide an array of options beyond staple crops, ensuring a varied and nutritious diet for individuals and communities. By cultivating a wide range of horticultural crops, farmers can reduce dependence on a single crop and mitigate the risks associated with crop failures or climate-related challenges. Shukla *et al.* (2016a) nicely explained powdery mildew disease in mango *vis-à-vis* weather related models for effective prediction and control management. Even, dynamic changes in weather based pan evaporation and reference evapotranspiration also indicated to modify moisture conservation strategy in Mango cv. Langra for better productivity (Adak *et al.,* 2019). These crops contribute to the beauty and aesthetics of landscapes, gardens, and public spaces. Flowers, ornamental plants, and landscaping elements enhance the visual appeal of surroundings, provide a sense of well-being, and offer recreational opportunities. Horticulture also plays a significant role in the tourism industry, attracting visitors to gardens, parks, and botanical attractions. The horticultural crops are deeply embedded in cultural practices and traditions worldwide. They hold symbolic and ceremonial importance, forming an integral part of festivals, rituals, and social gatherings. Horticultural activities, such as gardening and community farming, promote social cohesion, enhance mental well-being, and provide opportunities for intergenerational bonding and learning. Many horticulture crops have high export value due to their quality, uniqueness, and demand in international markets. Exporting horticultural produce generates foreign exchange earnings, stimulates rural economies, and supports agricultural trade. Overall, horticultural crops play a pivotal role in sustaining livelihoods, improving nutrition, preserving the environment, and enriching human experiences. They contribute to both local and global food systems, emphasizing the importance of their cultivation, conservation, and sustainable production.

Fig. 1: Distribution of vegetable seeds and vegetable production

Role of Farm Women In Dissemination of Advanced Farmers Friendly Technologies

Farm women play a significant role in the production of horticultural crops. Their contributions span various aspects of farming, management, and community engagement. They actively participates in the management of horticultural farms. They contribute to decision-making processes, crop planning and resource allocation. Their involvement ensures a holistic approach to farming practices, considering factors such as crop rotation, pest and disease management, soil health, and water conservation (Pankaj and Choudhary, 2019). In tribal areas, women farmers often take the lead in selecting and preserving seeds for horticultural crops. They possess valuable knowledge about local seed varieties, cultivation techniques, and seasonal planting practices (Fig. 1). Women's expertise in seed selection ensures the preservation of traditional crop varieties and promotes agro-biodiversity. Farm women are actively involved in the cultivation and care of horticultural crops. They participate in activities such as preparing the soil, sowing seeds, transplanting seedlings, applying fertilizers and pesticides, and tending to the crops throughout their growth stages (Barrios *et al.*, 2006). In hilly areas, women often employ traditional knowledge and sustainable practices to nurture and protect trees/vegetables. There is a need to disseminate zinc nutrition technology impacting on fruit yield and quality in fruits for the betterment of produce (Adak *et al.*, 2018). Even, diseases controls like anthracnose with humid thermal index (Shukla *et al.*, 2016b) and leaf webber and its seasonal incidence *vis-à-vis* weather based forecasting model (Gundappa *et al.*, 2016b) are essentially provides decision support in managing the crops. Farm women frequently manage irrigation and water-related activities on farms. They ensure the efficient use of water resources by employing techniques like drip irrigation, mulching, and rainwater harvesting (Kumar and Joshi, 2018). Since, soil moisture variability had significant impacts and tree water dynamics affects productivity in orchards (Adak *et al.*, 2020), Women's involvement in precise water management contributes to sustainable agriculture and helps address water scarcity challenges. Moreover, improvement of soil health condition is also an issue and farmers often seriously engaged in improving the soil health condition. This is particularly relevant for low fertile soils where soil application of fertilizers (both chemical and organic) and biofertilizers could offer positive response (Shukla *et al.*, 2014). Even, mulching, micronutrient spray and fertigation system if adopted seriously at farmers' field level, may boon income of growers (Barman *et al.*, 2017). Thus the need for efficient irrigation and fertigation scheduling even in ultra high density planting system may revolutionise yield and economics of fruits (Kumawat *et al.*, 2017). In addition to on-farm operations, farm women are actively involved in harvesting and post-harvest management. They are involved in activities such as picking ripe produce, sorting, grading, cleaning, and packaging the harvested crops. Women's attention to detail and knowledge of quality standards contribute to reducing post-harvest losses and maintaining the freshness and market value of the produce. Farm women often engage in value addition activities to enhance the economic viability of horticultural crops. They may process fruits and vegetables into various products such as jams, pickles, sauces, or dried fruits. Thus, women's involvement in value addition increases the shelf life of produce, creates new market opportunities, and adds value to the farm products. Women farmers also actively participate in marketing and sales activities related to horticultural crops. They may sell the produce directly in local markets, set up farm stands, or participate in farmers' markets. Additionally, women are increasingly utilizing online platforms and social media to promote and sell their farm products, contributing to the development of alternative marketing channels. Additionally, farm women often participate in knowledge sharing initiatives and community engagement activities related to horticultural production. They actively contribute to farmers' groups, self-help organizations, and community networks. Women's involvement in such platforms allows for the exchange of experiences, ideas, and best practices, leading to continuous learning and the adoption of sustainable farming techniques. Henceforth, women significant contribution towards dissemination of improved farmers' friendly technologies should be praiseworthy.

Fig. 2: Women's involvement in tree protection and packaging of harvested fruits

Women Empowerment and Contribution Towards Sustainability

In the era of entrepreneurship development, women empowerments are very much crucial in attaining the sustainable developmental goals. The role of farm women in the production of horticultural crops is essential for achieving sustainable agriculture, ensuring food security and strengthening rural economies. Their knowledge, skills and active participation contribute to the success and resilience of horticultural ecosystems. Thus, farm women play pivotal role in the sustainable production of horticultural crops. Their contributions are multi-faceted and encompass various aspects of farming, management, and community engagement. The role farm women play in sustainable production of horticulture crops is summarized under following categories:

1. **Organic Farming Practices:** Strategies for adoption of organic farming methods, women farmers often advocate their communities and also put in practice. They minimize the use of chemical inputs, promote natural pest control methods, and focus on enhancing soil fertility through composting, green manure, and crop rotation. By adopting organic practices, women farmers contribute to sustainable horticultural production by minimizing environmental pollution, protecting biodiversity, and ensuring food safety. The general overview of the role of women in organic horticulture production based on existing research indicate that women have played a crucial role in promoting and implementing sustainable farming practices in organic horticulture (Anand and Bahuguna, 2018). Women tend to have a deep understanding of local ecosystems and possess knowledge about traditional farming

techniques, which they adapt and integrate into organic farming methods. Their contributions include organic pest management, soil fertility management, biodiversity conservation and the use of traditional seeds (Gisore *et al.*, 2018). Women's involvement in organic horticulture has provided them with opportunities for empowerment and leadership roles. Studies have shown that women often take on decision-making responsibilities, organize farmer groups, and participate in community initiatives related to organic farming (Kamau *et al.*, 2019). Women's contributions in organic horticulture have had positive impacts on food security and nutrition. Organic horticulture production, led by women, has increased the availability of fresh and nutritious food at the household and community levels (Mwongera *et al.*, 2018). Their focus on crop diversification and the cultivation of traditional and indigenous crops has helped improve dietary diversity and address malnutrition issues (van der Merwe *et al.*, 2016).

2. **Seed Preservation and Conservation of Agro-biodiversity**: Women farmers are often the custodians of traditional seed varieties. They play a vital role in preserving and promoting local seed diversity, which contributes to the resilience of horticultural systems. They often play a crucial role in collecting and preserving traditional and locally adapted seeds. They possess knowledge about seed selection, storage techniques. Women farmers save and exchange seeds, maintain seed banks, and pass on their knowledge of local varieties to future generations, thus safeguarding genetic diversity and promoting sustainable farming practices (Badstue *et al.*, 2007). Farm women are actively involved in seed production, including seed sowing, nurturing crops, and ensuring seed quality. They engage in activities such as pollination, rouging (removing off-type plants), and maintaining seed purity. Women's active participation in on-farm seed production contributes to maintaining diverse and resilient seed systems (Mburu *et al.*, 2018). Women frequently engage in informal seed sharing networks within their communities. They often share seeds, knowledge, and farming practices, contributing to the conservation and exchange of crop biodiversity. These networks facilitate the maintenance and diversity of local seed varieties of vegetables, underutilized fruits, and medicinal plants. In some cases, women take on entrepreneurial roles in seed-related activities. They may engage in small-scale seed businesses, selling seeds from their farms or participating in community seed banks. These initiatives empower women economically and strengthen local seed systems. Women's participation in seed preservation and conservation extends beyond

on-farm activities. They actively engage in advocating for policies that support the conservation of crop biodiversity and the recognition of traditional seed systems. Women's involvements in decision-making processes related to seed policies helps ensure their perspectives and needs are considered.

3. **Sustainable Water Management:** Women frequently take responsibility for water resource management especially in regions where water scarcity is a challenge. Farm women play a significant role in implementing water conservation techniques and efficient irrigation practices in horticultural crops. Their involvement ranges from on-farm activities such as water harvesting, mulching, and drip irrigation to community-level initiatives promoting sustainable water management. Women's participation in water governance and decision-making processes at various levels is crucial for addressing gender-specific water needs, ensuring equity, and promoting sustainable water management practices (Zwarteveen, *et al.*, 2018). Women are often knowledge bearers and agents of change in sustainable water management. Farm women engage in various water-related livelihood activities associated with horticultural crops, such as nursery management, post-harvest processing, and value addition. Their participation in these activities contributes to the overall sustainability of water use and crop production (Mukherji and Suhardiman, 2018). They play a critical role in sharing traditional knowledge, best practices, and innovative approaches for water conservation in horticultural crops. Capacity-building initiatives targeting women can enhance their skills and knowledge in sustainable water management (Manglitz *et al.*, 2019).

4. **Soil Conservation and Nutrient Management:** Women farmers focus on maintaining soil health and fertility through sustainable soil conservation and nutrient management practices. Women are actively engaged in implementing soil conservation measures on farms. They contribute to activities such as contour farming, terracing, mulching, and agroforestry, which help prevent soil erosion, improve soil structure, and conserve soil moisture (Franzel *et al.*, 2005). Women farmers play a crucial role in nutrient management by managing the organic waste on farms and converting it into compost. They collect crop residues, farm and household waste, and livestock manure, and turn them into nutrient-rich compost, thereby enhancing soil fertility and organic matter content (Akter and Sobhan, 2019). Women actively participate in agroforestry practices and tree planting activities. They contribute to the establishment and maintenance of tree-based systems on farms, which help improve soil fertility, prevent erosion, and enhance nutrient cycling

(Kiptot and Franzel, 2012). They employ techniques like cover cropping, crop rotation, green manuring and addition of organic amendments to improve soil structure, enhance water-holding capacity, and reduce soil erosion. The role of green manuring and nutrient management *vis-à-vis* soil properties and sustainability of fruit orchard needs to be disseminated across stakeholders in order to adoption at local level (Kumar *et al.*, 2017). Henceforth, women's understanding of the importance of soil health contributes to the long-term sustainability of horticultural crops.

5. **Integrated Pest Management (IPM):** Women farmers often embrace integrated pest management approaches in horticultural production (Fig. 2). They use biological controls, crop rotation, trap crops, and natural pest repellents to manage pests and diseases sustainably. Women farmers actively participate in monitoring pest and disease outbreaks in horticultural crops. They observed plants for signs of pest and disease damage to identify the symptoms (Sarapulova, *et al.*, 2020). Farm women often possess traditional knowledge and practices for managing pests and diseases in horticultural crops. They use cultural practices such as companion planting, intercropping, and trap cropping to deter pests and reduce disease incidence (Rao and Maurya, 2019). They employ biological pest control measures in horticultural crops by rearing and releasing beneficial insects, such as predatory mites or parasitic wasps, to control pest populations naturally (Karanja *et al.*, 2018). Thrips is a serious menace in tropics and subtropics; therefore its dynamics *vis-à-vis* assessment using humid thermal ratio needs to be put in practices in order to safeguard the crop (Gundappa *et al.*, 2016a). By minimizing the reliance on synthetic pesticides, women farmers protect beneficial insects, pollinators, and natural ecosystems while ensuring crop health and productivity.
6. **Value Addition and Post-Harvest Management:** Farm women also play a significant role in value addition of vegetable crops, contributing to the processing, preservation, and marketing of vegetable-based products. They are involved in activities like sorting, grading, packaging, and processing of produce. In fruit crops, farm women are actively engaged in processing and preserving fruit crops to increase their shelf life and create value-added products. They are involved in activities such as making jams, jellies, pickles, dried fruits, juices, and fruit-based sauces, using traditional recipes and innovative techniques (Maikhuri *et al.*, 2021). Farm women often engage in value addition activities to enhance the economic viability of horticultural crops. They are involved in activities such as canning, pickling, drying, fermenting, and making

sauces, soups, and ready-to-eat meals using various vegetables (Chandra *et al.*, 2019). Farm women contribute to diversifying vegetable-based products by introducing innovative recipes, flavors, and packaging. They utilize traditional knowledge and techniques to create unique value-added products that cater to specific consumer preferences (Tripathi and Singh, 2018). They are actively engaged in flower arrangement and bouquet making, utilizing their creativity and artistic skills to create attractive and marketable floral designs. They arrange flowers in various patterns and styles, catering to different occasions and customer preferences. They make items such as garlands, wreaths, floral accessories, and home decor products, drying and preservation of flowers to extend their shelf life and create value-added products like crafts, potpourri, and floral arrangements, thus adding value to the floral industry (Mahanta *et al.*, 2020). By adding value to the harvested crops, women contribute to reducing post-harvest losses and ensuring better market access.

7. **Community Engagement and Empowerment:** Women farmers often form or participate in self-help groups, cooperatives, and community-based organizations. Digital tools like Whats App, mobile apps and web based system, decision support system etc were highly beneficial to them for easy communication and diffusion of technologies for economic benefits. These platforms provide opportunities for women to access resources, information, and market networks. Through collective action, women farmers can advocate for their needs, improve their bargaining power, and contribute to the sustainable development of their communities. These groups often facilitate training programs on sustainable farming practices, financial literacy, and entrepreneurship. By actively participating in such groups, women gain knowledge, build networks, and collectively work towards sustainable horticultural production.

Farm women's involvement in horticultural production contributes to their economic empowerment and the overall well-being of their families. Income generated from sustainable farming practices enhances livelihoods, reduces dependence on external sources and promotes self-reliance. This empowerment has positive social and economic ripple effects in rural communities. It is important to recognize and support the significant contributions of farm women in the sustainable production of horticultural crops. Their knowledge, skills, and active participation are critical for fostering environmentally friendly farming practices and achieving sustainable food systems. The active involvement of women farmers in sustainable horticultural production is crucial for promoting environmentally friendly farming practices, ensuring food security and achieving sustainable development goals. Recognizing and supporting the role of farm women in sustainable horticulture production is crucial for achieving long-term agricultural sustainability. Policy makers, researchers and practitioners should prioritize the inclusion of women in decision-making processes, provide access to resources, training, and market opportunities, and create an enabling environment that promotes their active involvement. By harnessing the potential of farm women, sustainable horticulture production systems can be fostered. Overall, farm women are key agents of change in promoting sustainable production of horticulture crops and their contributions are essential for building resilient and inclusive agricultural systems.

References

Adak, T., Babu, N. and Pandey, G. (2020). Soil Moisture Variability and Tree Water Status vis-à-vis Productivity in Fruit Orchards as Estimated by Unmanned Aerial Vehicle, Drones, RADAR Technologies. Journal of Agricultural Physics 20(2): 185-190.

Adak, T., Shukla, K., Kumar, K. and Singh, V. K. (2018). Effect of different Zinc nutrition on fruit yield and quality in Mango under Lucknow condition. In: Proceedings of International Mango conference held during 8-11th May 2018 at Vengurle, India, pp. 95-96.

Adak, T., Singh, V. K., Kumar, K. and Pandey, G. (2019). Moisture Conservation and Productivity in Mango cv. Langra using Drip Fertigation. In: Soil and Water Conservation Bulletin, Mishra, P. K., Sharma, N. K., Kumar, A., Mandal, D., Kaushal, R., Kumar, G., Roy, T. and Dogra, P. (Eds), Indian Association of Soil and Water Conservationists, Dehradun, Uttarakhand, 4: 77-80.

Adak, T. and Pandey, G. (2018). Soil Management of Horticultural Crops for Food and Nutritional Security of Farm Women. In: Training Manual on Model Training Course Food and Nutritional security of farm women through horticulture based interventions, during 13-20 November held at ICAR-CISH Lucknow, India, pp. 111-114.

Akter, S. and Sobhan, F. A. (2019). Empowering women through composting: A case study from Bangladesh. Journal of Gender, Agriculture and Food Security 4(1): 1-21.

Anand, P. and Bahuguna, V. K. (2018). Gender participation in organic farming: A study of Uttarakhand, India. Journal of Rural Studies 59: 98-106.

Babu, N., Adak, T. and Kumar, A. (2022). Horticulture intervention for improving productivity and profitability of small and marginal farmers. Journal of Natural Resource Conservation and Management 3(1): 49-57

Badstue, L., Bellon, M. R., Berthaud, J., Ramírez, A., Flores, D. and Juárez, X. (2007). Examining the role of collective action in an informal seed system: A case study from the Central Valleys of Oaxaca, Mexico. World Development 35(5): 851-865.

Barman, P., Srivastava, K. K. and Adak, T. (2017). Response of guava cv. Lalit to plastic mulch, Fertigation and micro-nutrient spray under raised bed. Indian Journal of Agriculture and Allied Sciences 3(4): 1-8.

Barrios, E., Delve, R. J. and Bekunda, M. A. (2006). Using manure as a resource: Sustainable nutrient management in smallholder agroecosystems. In B. A. Kratochvil, A. D. Zanin, A. J. Hunter, and D. A. Dyck (Eds.), Nutrient management in agricultural watersheds: A wetland solution (pp. 107-120). Springer.

Chandra, A. et al. (2019). Role of women in value addition of horticultural produce: A case study of processed vegetable products in Uttarakhand, India. Indian Journal of Agricultural Marketing 33(2): 167-177.

Franzel, S., Denning, G. and Lillesø, J. P. B. (2005). Women in extension: Building bridges to sustainable forestry. In C. Haverhals and B. Kerven (Eds.), Women and sustainable agriculture: Interviews with 14 agents of change (pp. 11-29). IIRR, The Netherlands.

Gisore, N., Mairura, F., Kahi, E. and Mburu, J. (2018). Gender participation in organic vegetable farming in Kenya: Implications for livelihoods improvement. Journal of Gender, Agriculture and Food Security 3(1): 70-86.

Gundappa, Adak, T. and Shukla, P. K. (2016a). Humid Thermal Ratio as a tool to assess mango thrips dynamics under subtropical climatic condition. Journal of Environ Biology 37:1239-1245.

Gundappa, Adak, T. and Shukla, P. K. (2016b). Seasonal incidence and weather based forecasting model for mango leaf webber under subtropical conditions. Current Advances in Agricultural Sciences, 8(1): 96-99.

Gundappa, Adak, T. and Shukla, P. K. (2016c). Seasonal dynamics of mango hoppers and their management under subtropics. GERF Bull Biosc, 7(1):6-9.

Kamau, W., Karanja, D. and Mugatha, S. M. (2019). Determinants of women's participation in organic horticulture production in Kenya: The case of Ruiru Sub-County. Journal of Agricultural Science 11(11): 253-265.

Karanja, D., Njenga, M. and Ochieng, R. (2018). Gender roles in managing the western flower thrips (Frankliniella occidentalis) in French beans among small-scale farmers in Kenya. Journal of Gender, Agriculture and Food Security 3(2): 48-64.

Kiptot, E. and Franzel, S. (2012). Gender and agroforestry in Africa: A review of women's participation. Agroforestry Systems 84(2): 181-196.

Kumar, K., Adak, T. and Singh, V. K. (2017). Green manuring and nutrient management impacting soil properties and sustainability of mango orchard. Journal of Soil and Water Conservation 16(1): 72-78.

Kumar, M. and Joshi, P. K. (2018). Role of women in water management for agriculture: Evidence from India. International Journal of Sustainable Development & World Ecology 25(6): 529-539.

Kumawat, K. L., Sarolia, D. K., Kaushik, R. A. and Jodha, A. S. (2017). Effect of irrigation and fertigation scheduling on growth, flowering, yield and economics of guava cv. Lalit under ultra high density planting system. Indian Journal of Horticulture 74(3): 362-368.

Mahanta, S. et al. (2020). Women's role in floriculture-based entrepreneurship development in Assam, India. International Journal of Agriculture, Environment and Biotechnology 13(6): 1099-1106.

Maikhuri, R. K. et al. (2021). Women-led innovations and entrepreneurship in value-added processing of horticultural crops: A study from Uttarakhand, India. Journal of Ethnic Foods 8: 49-60.

Shukla, P. K., Adak, T. and Gundappa. (2016a). Weather based models in predicting Powdery mildew disease for Malihabad mango belt. J. Eco-Friendly Agriculture, 11(2):142-148.

Shukla, P. K., Adak, T. and Gundappa. (2016b). Relationship between humid thermal index and dynamics of anthracnose disease of mango. In: Proceedings of 6th International conference on Plant, pathogens and people held during Feb. 23-27, 2016 at NASC complex, New Delhi, India. pp. 52-53.

Verma, H. C. and Adak, T. (2021). Role of digital tools to foster economic prosperity of fruit growers. In: 2nd International Web-Conference on "Smart Agriculture for Resource Conservation and Ecological Stability" organized by ANRCM during October 29 to 31, 2021. p 3.

Manglitz, A., Lautze, J. and Giordano, M. (2019). Building gender capacity in the water sector: Key principles and strategies. Water International 44(7): 689-706.

Mburu, J., Gichimu, B. M. and Mutitu, E. W. (2018). Women and seed systems in sub-Saharan Africa: A review of constraints and opportunities for enhancing women's participation. International Journal of Biodiversity and Conservation 10(9): 386-396.

Mukherji, A. and Suhardiman, D. (2018). Engendering irrigation organizations in Nepal and Vietnam. Water Alternatives 11(2): 526-548.

Mwongera, C., Shisanya, C., Njiru, E. and Midega, C. A. O. (2018). Factors influencing women participation in organic vegetable farming: A case of Kirinyaga County, Kenya. Journal of Biology, Agriculture and Healthcare 8(10): 125-139.

Pankaj, S. and Choudhary, K. (2019). Empowering women through training in water resource management: An analysis from eastern India. Journal of Natural Resources Policy Research 11(4): 287-302.

Rao, V. P. and Maurya, D. M. (2019). Traditional pest management practices followed by tribal farmers in Rajasthan. Journal of Pharmacognosy and Phytochemistry 8(1): 1929-1933.

Sarapulova, S. V. et al. (2020). Socio-economic aspects of managing agricultural pests and diseases in Tajikistan: A gender perspective. Agriculture 10(9): 368.

Shukla, S. K., Adak, T., Singha, A., Kumar, K., Singh, V. K. and Singh, A. (2014). Response of guava trees (Psidium guajava) to soil applications of mineral and organic fertilisers and biofertilisers under conditions of low fertile soil. Journal of Horticultural Research 22(2): 105- 114.

Tripathi, M. and Singh, R. (2018). Women entrepreneurship in vegetable processing and value addition: A case study of Lucknow, India. International Journal of Chemical Studies 6(3): 140-145.

van der Merwe, J. D., Cloete, P. C. and van der Hoeven, M. (2016). Promoting food security through indigenous and traditional food crops. Agroecology and Sustainable Food Systems 40(8): 830-847.

Zwarteveen, M., Ahmed, S., Akter, S. and Gautam, S. (2018). Doing gender justice in water governance and water interventions. Water Alternatives 11(1): 1-20.

Authors' Index

Index